Evolution
&
Ethics

T. H. HUXLEY IN 1893

(Photo by Mayall, 1893)

Evolution
&
Ethics

T. H. Huxley's

Evolution and Ethics

With New Essays on
Its Victorian and
Sociobiological
Context

◆

JAMES PARADIS
GEORGE C. WILLIAMS

Princeton University Press
Princeton, New Jersey

Copyright © 1989 by Princeton University Press
Published by Princeton University Press,
41 William Street, Princeton, New Jersey 08540
In the United Kingdom: Princeton University Press,
Guildford, Surrey

All Rights Reserved

Library of Congress Cataloging-in-Publication Data
Paradis, James G., 1942–
Evolution and ethics.
Includes bibliographies and index.
1. Huxley, Thomas Henry, 1825–1895. Evolution and
ethics. 2. Ethics, Evolutionary. I. Huxley, Thomas
Henry, 1825–1895. Evolution and ethics. 1894.
II. Williams, George C. (George Christopher), 1926–
III. Title.   BJ1311.H83P37 1989 171'.7 88-32530
ISBN 0-691-08535-8 (alk. paper)
ISBN 0-691-02423-5 (pbk.)

This book has been composed in Linotron Baskerville

Clothbound editions of Princeton University Press books
are printed on acid-free paper, and binding materials are
chosen for strength and durability. Paperbacks, although satisfactory
for personal collections, are not usually suitable for library rebinding

Printed in the United States of America
by Princeton University Press, Princeton, New Jersey

Contents

Preface

THOMAS HUXLEY'S Romanes Lecture, "Evolution and Ethics," and its introductory "Prolegomena," written during the socially chaotic Victorian 1890s, are classics of fin-de-siècle literature. Like most classics, they treat themes that continue to have immense human interest. Huxley's views that the human psyche is at war with itself, that humans are alienated in a cosmos with no special sensitivity to their needs, that moral societies are of necessity in profound revolt against the natural conditions of their existence—concepts that all fit within the framework of an evolutionary world view—remain as controversial today as they were when he first proposed them.

Huxley's Promethean call to his fellows to pit the powers of the industrial revolution against the forces of nature brings up questions about the human relationship with the environment that are every bit as important and troubling today as they were nearly one hundred years ago. Modern fields of inquiry continue to ask these and many related questions, but Huxley's essays show the origins of the revolt against nature in a Victorian synthesis that reflects upon arguments long associated with Adam Smith, Thomas Malthus, Charles Darwin, Herbert Spencer, Edmund Tylor, Karl Marx, Walter Bagehot, and John Stuart Mill. In Huxley's two essays, we have a mirror of Victorian naturalistic thought, as well as a behavioral argument of great interest to the contemporary evolutionary biologist.

In this volume, we make available Huxley's two essays—referred to jointly as *Evolution and Ethics*—and examine Huxley's position first in the context of nineteenth-century cultural history, and then in respect to twentieth-century evolutionary thought. The introductory essay traces Huxley's ideas to their eighteenth-century origins while considering the uses Huxley made of Victorian anthropological, historical, and evolutionary thought as he framed what to him was the key social statement of his career. Huxley's two essays, the introductory "Prolegomena" and "Evolution and Ethics," appear in the first and authoritative Macmillan edition of 1893 and are followed by new notes and translations of the Greek and Latin quotations. The concluding essay, which has appeared in abbreviated form in *Zygon*, volume 23, number 4, examines Huxley's views in relation to modern biological thought. An appendix on the history and criticism of *Evolution and Ethics* has been prepared by James Paradis. The bibliographical entries are listed in two sections. The first lists the works cited in "T. H. Huxley's *Evolution and Ethics* in its Victorian Context," as well as in the appendix. The second lists the works cited in "A Sociobiological Expansion of *Evolution and Ethics*." In thus moving from Victorian cultural history to contemporary biology, we hope to provide new insights into the origins and persistence of biosocial dilemmas that have considerable contemporary significance.

The authors are grateful for assistance by many individuals whose criticism and editorial suggestions helped to broaden the scope and improve the arguments of the volume. James Paradis would like to thank Bernard Lightman, Kenneth Manning, and the manuscript reviewers for Princeton University Press for their detailed and helpful suggestions on various portions of the introductory essay. He would also like to thank Edward Barrett for his help in translating the Greek and Latin quotations in T. H. Huxley's two essays. For their support, he would like to thank Judy, Emily and Rosalind. George Williams found time for the writing and library work largely as the result of a 1981–1982 fellowship at the Center for Advanced Study in the Behavioral Sciences at Stanford, California. His manuscript was much improved by comments from Douglas J. Futuyma, Sarah Blaffer Hrdy, Paul W. Sherman, Doris C. Williams, and David Sloan Wilson. Judith May was a persistent and prescient editor for Princeton University Press during the early stages of this volume until her departure for England. Emily Wilkinson took over and provided valuable guidance and Julie Marvin gave many helpful editorial suggestions on the manuscript. In addition, several archivists and librarians at the following institutions helped to locate source materials and clarify bibliographical details: the British Museum, the Tozzer Library and the Museum of Natural History at Harvard University, and the Barker Library at M.I.T.

Evolution
&
Ethics

Evolution and Ethics
in Its Victorian Context

♦

JAMES PARADIS

IN THE SUMMER of 1892, three years before his death, an ailing T. H. Huxley wrote the celebrated lecture "Evolution and Ethics," which he delivered at Oxford University the afternoon of May 18, 1893. The lecture, together with the "Prolegomena," an introductory essay completed in June of 1894, set traditional humanistic values in direct conflict with the physical realities revealed by nineteenth-century science. The forces of nature, seen by Huxley in terms of powerful material and instinctual laws, were poised, he now argued, against civilization and the future of humanity.

Huxley built his two essays on a domestic foundation, using a wealth of autobiographical themes and images. The struggle against odds, the need for self-restraint, the summoning of courage to strip off the veil of nature and remove the garment of make-believe, all lifelong personal themes, were transformed in the manner of Montaigne into dimensions of the human condition. Willingly or not, we are all controversialists in a transitory universe:

> The more we learn of the nature of things, the more evident is it that what we call rest is only unperceived activity; that seeming peace is silent but strenuous battle. In every part, at every moment, the state of the cosmos is the expression of a transitory adjustment of contending forces; a scene of strife, in which all the combatants fall in turn. (p. 49)

Here, in the depths of the aging Huxley's dynamism, lies a symbol of the Victorian age itself, like Huxley, tottering at the edge of the abyss its science had so irrevocably revealed. This sense of peril is reflected in the images of the colony that Huxley used in his two essays to characterize the polity, civilization, and England itself.

3

The prolific garden imagery of the "Prolegomena" had an origin in Huxley's recent occupation of a small house, Hodeslea, he had built for his retirement on the windswept chalk downs above the sea at Eastbourne, a fitting outpost for the author of "On a Piece of Chalk" (1868). Gardening had become, along with controversy, his occupation, leading him to cite Voltaire's Candide as his own—and civilized man's—proper spiritual mentor; "Cultivons notre jardin" (Huxley 1892, 564). The wall of civilization inscribing the artificial world that Huxley described in the "Prolegomena" had its physical analogue in his own garden wall; he wrote that out his study window beyond his garden he could see the "state of nature" to which all was destined eventually to return. These personal images gave Huxley's essays the intensity and clarity that moved Leslie Stephen at a meeting of the Huxley Memorial Committee to place Huxley among the greatest of English prose artists (Stephen 1895).

It is well to recall that Huxley wrote *Evolution and Ethics* at a time when it was not clear what role, if any, the forces of science had to play in the framing of social policy—a peculiarly modern ambiguity. As Burrow (1966), Young (1985), and Stocking (1987) have shown, writers like Herbert Spencer and Walter Bagehot had broadly integrated popular biology and contemporary anthropology with social argument, often through the vehicles of comparison and metaphor. At the same time, the natural sciences had undermined many traditional cultural assumptions concerning the natural order, causing politicians, clergy, and intellectuals alike to reexamine the sources of social, intellectual, and moral authority (Manier 1978; Moore 1979).

This search for authority was not merely academic. Voting suffrage, extended first to urban laborers and then to miners and farm laborers in the Second and Third Reform Bills of 1867 and 1884, had dramatically expanded the English electorate and redistributed political power. In addition, as Steadman Jones has demonstrated (1971), London in the 1880s and 1890s was a scene of cyclic economic depression attended by great poverty and social unrest. If the century had been "predominantly a history . . . of emancipation, political, economical, social, moral, intellectual," as William Gladstone observed soon after resigning his fourth premiership in March of 1893, that same emancipation, he added, was opening "a

period possibly of yet greater moral dangers; certainly a great ordeal for those classes which are now becoming largely conscious of
power, and never heretofore subjected to its deteriorating influences" (Reid 1899, 2:732). This sense that new social forces had
been unleased was felt throughout British society.

Huxley, Gladstone, and their aging contemporaries saw the social
speculation of the time as competing for the hearts and minds of
the emerging voters. A great variety of radical political movements
had asserted themselves, each with its unique program and future:
Henry Hyndman's Social Democratic Federation, General William
Booth's Salvation Army, Fabian socialism, Henry George's Single
Land Tax movement, Francis Galton and Karl Pearson and the Eugenics movement. Books published in the years leading up to Huxley's *Evolution and Ethics* included Arnold Toynbee's *The Industrial
Revolution* (1884), the English edition of Friedrich Engels's *The Condition of the Working-Class in England* (1892), the second edition of
Galton's *Hereditary Genius* (1892), George's *Progress and Poverty*
(1877–79), General Booth's *In Darkest England* (1890), Edward Bel-
lamy's *Looking Backward: 1888–2000* (1888), and Spencer's *Principles of Ethics* (1892–93), not to mention William Gladstone's *Impregnable Rock of Holy Scripture* (1890)—all works considering the values
and measures that might secure the future of society. This bookish
activism, avidly followed by Huxley, had its social counterpart in
the atmosphere of conflict in London in the 1880s and 1890s, the
years of economic instability that witnessed the 1886 Riot, Bloody
Sunday (1887), and the Dock Strike (1889), as well as innumerable
lesser social tremors and the vast political controversy over Irish
Home Rule.

In the most important study to date of the social context of Huxley's *Evolution and Ethics*, Michael Helfand has rightly argued that
Huxley, struck by the radical political agitation of the 1880s and
1890s, had an overt political motivation for writing the two essays.
Helfand, however, has constructed an elaborate texture of "hidden" intention that makes Huxley over into a repressive partisan of
the status quo. Against formidable intellectual evidence to the contrary, Helfand is obliged to argue an externalist thesis that Huxley's
"real" view of contemporary society was actually a "disguised"
Spencerian laissez-faire dynamic of struggle for existence (Helfand

5

1977, 160–61). Huxley's motivation under this interpretation was to defend an entrenched middle class against the rising working classes and their anti-imperialist champions, Henry George and Alfred Wallace. This conspiratorial thesis of class conflict, aside from giving more political unity and credibility to Huxley's opponents than is perhaps justified, is too reductive for an individual of Huxley's experience and complexity.

To be sure, Huxley, formerly a middle-class administrator in London, was dismayed by the social volatility of contemporary London, which seemed to him to threaten the institutions that protected society from anarchy. A recognized reformer, Huxley was also a social realist who had spent innumerable hours on public commissions and institutional committees. As a practitioner familiar with the demands of social change, he saw social stability as the condition of social amelioration. Revolution in the cause of speculative political theory went against Huxley's most fundamental convictions and training. His opposition to groups of radicals, from anarchists and Salvation Army members to eugenicists, was a protest against the exotic utopian character of their programs. His objections in earlier essays to the sweeping anti-Malthusianism of George's *Progress and Poverty* had been much more elaborately articulated in Toynbee's widely-known criticisms of George in *The Industrial Revolution*.

In an essay on Charles Darwin's theory of natural selection, Huxley had once observed, "The struggle for existence holds as much in the intellectual as in the physical world. A theory is a species of thinking, and its right to exist is coextensive with its power of resisting extinction by its rivals" (1893–94 2:229). The 1880s and 1890s furnished graphic evidence of this conceptual struggle, which Huxley took up by writing on social issues from his own professional and personal perspective. Given populist politics, given the power of the later industrial revolution, and given the intellectual and social implications of contemporary evolutionary theory, anthropology, and biblical criticism, how, Huxley asked, are we to affirm an ethical principle for human society? His answer was to reassert, in a series of essays culminating with *Evolution and Ethics*, a modified Malthusian argument based on the natural inequality between the forces of population and production.

Huxley wrote the two essays of *Evolution and Ethics* in imitation of

Malthus for precisely the reasons that Malthus had written *An Essay on the Principle of Population* (1798)—to refute, by reference to natural material and instinctual constraints, the romantic a priori arguments of the social idealists. Like many fin-de-siècle Victorians, Huxley feared that violent revolution might be at hand (Jones 1971, 290). In an introduction to the English translation of Théodore Rocquain's anticlerical history of the French Revolution, *The Revolutionary Spirit*, he had written:

> The grave political and social problems which press for solution at the present day are the same as those which offered themselves a hundred years ago. Moreover, the a priori method of the *Philosophes*, who, ignoring the conditions of scientific method, settled the most difficult problems of practical politics by fine-drawn deductions from axiomatic assumptions about natural rights, is as much in favor at the end of the nineteenth, as it was in the latter half of the eighteenth century.
> (Huxley 1891, vii–viii)

But Huxley's reservations about radical political theory did not make him an advocate of social status quo or a partisan of Spencerian laissez-faire. Unlike Malthus, he had witnessed the material powers of the industrial revolution. If human society were governed by an evolutionary dynamic, that society's abilities to transform its conditions of existence might go far in neutralizing the formidable natural constraints on the individual. This theme of social transformation is the burden of *Evolution and Ethics*.

In education and human artifice, Huxley found two principles of transformation by which the so-called state of art supplanted the strictly evolutionary dynamic of the state of nature. The key to Huxley's political philosophy, if we may call it that, is education. Although this philosophy is not elaborated in the two essays of *Evolution and Ethics*, it is spelled out in detail in such essays as "A Liberal Education and Where to Find It" (1868), "The School Boards: What They Can Do, and What They May Do" (1870), "Universities: Actual and Ideal" (1874), and a dozen other essays on educational topics, collected in volume 3 of his *Essays*. Huxley wrote the classic nineteenth-century essays on science education, which he viewed as the means to transforming society, liberating individuals, and establishing a free market of merit and ideas—a theme that was to figure

7

prominently in John Dewey's philosophy. This was for Huxley a profoundly autobiographical theme, reflecting upon his meteoric rise from Free Scholarship student at Charing Cross Hospital in 1842–1845 to president of the Royal Society from 1883–1885. Yet education, seen within the evolutionary dynamic, was also the social vehicle that transmitted human experience, the source of society's power to transform its surroundings (Noland 1964).

Huxley's Romanes Lecture and "Prolegomena," with all their political and autobiographical elements, were preeminently Huxley's final defense of Victorian naturalistic thought (Barton 1983). Anchored in eighteenth-century ethical categories derived from Malthus, Hume, and Hartley, the two essays invoked a Victorian anthropological perspective to argue that ethical behavior was part of a universal cultural dynamic that both depended upon physical and biological circumstances and sought to break free of them. In this dynamic, the biological and sociocultural implications of ethics were not easily reconciled. Physical nature, operating outside the compass of ethics, which was decidedly a human cultural artifice, could not furnish a norm for ethics. From the moralist's practical perspective, the cosmic background was not simply neutral, but embodied forces that effectively, although not intentionally, frustrated human ethical intent. It was antagonistic in the sense of tragedy, in which unknown forces within and without the individual bring forth chaos. The ethical individual was, therefore, knowingly or not, in revolt against the macrocosm. Human artifacts played a decisive role in this struggle, a role greatly magnified by the powers of the industrial revolution. With this reasoning, Huxley denied the romantic impulse to spiritualize the natural and pitted science and consciousness against nature and instinct. In these two major essays of his late career, Huxley achieved a plausible intellectual and social synthesis consistent with Victorian evolutionary, anthropological, and technological developments, a synthesis that explains his appeal and modernity as a Victorian man of letters.

Huxley and Eighteenth-Century Naturalistic Thought

In *Evolution and Ethics*, Huxley locates the historical origins of Victorian ethical naturalism in the diffuse speculative traditions of the previous century. Two of these traditions made natural process the a priori metaphor of morality (Lovejoy 1936, 289). The familiar literary neoclassicism of such figures as John Dryden, Joseph Addison and Henry Steele, and Alexander Pope, promoted as both social and aesthetic truth the old stoical injunction, *live according to nature* (p. 73). This aphorism's social validity is a matter to which Huxley's two essays continually return. A second, equally prominent, line of British ethical naturalism was established by the physicotheologians, who in the analogical tradition of the great natural philosophers John Ray, Robert Boyle, and Isaac Newton, celebrated the fruits of natural inquiry as evidence of the great cosmic order of the Creator. Within this tradition, which continued unabated through William Paley's *Natural Theology* (1802) to Robert Chambers's *Vestiges of the Natural History of Creation* (1844), physical nature, duly schematized and interpreted, furnished a value-laden social norm that proclaimed the very moral codes of established religion (Gillispie 1951; Hankins 1985). As both neoclassical and physicotheological traditions thus vigorously appealed to physical nature as the standard of reason and morality, so, too, did they certify the existing social order, all of which led Basil Willey to characterize them as philosophies of "cosmic Toryism" (1940, 34–36, 55).

Developing somewhat in counterpoint to this eighteenth-century literary naturalism, however, was a third, equally bold trend of viewing human behavior from the normative perspective of populations. Assuming unity of kind in causes and effects, Jeremy Bentham, David Hume, Adam Smith, and Thomas Malthus all deliberately softened the distinctions between humans and brute animals in order to fix the individual on a material grid whose abstract terms would support a new kind of social analysis. Adam Smith, for example, argued in his *Wealth of Nations* (1776) that any "species" of animal, including humans, "multiplies in proportion to the means of [its] existence," child mortality being one of the key factors of social adjustment (1776, 1:97). Smith extrapolated the systematic consequences of group behavior in the notion of an *economy*—a closed, material, self-adjusting system. This material view of humankind was central to Malthus's study of population. In *An Essay on the Principle of Population* (1798), Malthus derived from the material circumstances of both animals and humans laws, based on populations, that identified certain forces irresistibly influencing individual behavior, including moral behavior. These forms of analysis, which derived norms of behavior from representative circumstances, established a new methodological naturalism that found its ultimate Victorian expression in the population statistics of the Belgian astronomer Adolphe Quételet and the evolutionary speculation of Darwin (Ghiselin 1969, 49; Schweber 1977, 287–93; Ruse 1979, 145–46).

Huxley's *Evolution and Ethics*, appearing at the end of the nineteenth century, reflected back on the naturalisms that had contributed so considerably to the achievements of Darwin and his contemporaries, including Huxley himself. Malthus had stated in the clearest terms that the dilemma of an ethical society is biological. The Malthusian inequality between population and production—a secular analogue to fallen nature—remained the central ethical obstacle recognized by Huxley nearly a century later. From Hume, on the other hand, Huxley borrowed moral categories and arguments, especially the instinctual category of *sympathy*—similar, in some respects, to Christian grace—that he had originally established in his full-length critical monograph, *Hume* (1878). Finally, in Hartley's associationist psychology, Huxley found a precedent for a physiological explanation of moral development that suggested a material basis for the evolution of ethical behavior.

The driving energies of Huxley's *Evolution and Ethics* derive from the Malthusian "natural inequality" between population and production, which is to say between nature and humans. In both men's works, biological limits give rise to and yet threaten mind and society. Both Malthus and Huxley sought to expose the methodological weaknesses of contemporary a priori social speculators. Malthus, in the naturalistic language of his contemporary, the geologist James Hutton, cited the "constancy of the laws of nature and of effects and causes" in opposition to the "conjectures" of Condorcet and Godwin concerning the perfectibility of humans as organic and moral beings (1798, 126, 162). Huxley, likewise, appealed to natural process in order to expose the fallacies of "modern speculative optimism, with its perfectibility of species, reign of peace, and lion and lamb transformation scenes" (p. 78). His adversaries—laissez-faire social speculators such as Spencer and Petr Kropotkin, and social idealists such as George and General William Booth—might as easily have been Malthus's own targets.

For Malthus, the resources of nature were unequal to the needs of its progeny, which made nature impossible to accept as a social and ethical norm. From the Renaissance through the eighteenth century, the concept of the natural was richly invested with standards of order and harmony. "*Nature*," Lovejoy once remarked, "[was] the chief and most pregnant word in the terminology of all the normative provinces of thought in the West" (1948, 69). But Malthusian nature, as recent commentators have observed, with its "natural inequality of the two powers of population and production," constituted a physical and psychical environment of enduring conflict (Young 1985; Kohn 1980; Ospovat 1981). Such conflict led to a social dilemma of profound human proportions.

The Malthusian view thus ruptured the traditional myths and ideologies of the natural, which had for centuries found a worthy moral and artistic referent in the order and permanence of the natural world (Tillyard 1942; Willey 1940; Gillispie 1951). A destabilized physical nature of the kind Malthus had imagined in his *Essay on Population* had little capacity to furnish an ethical norm:

> Necessity, that imperious all pervading law of nature, restrains [all existence] within the prescribed bounds. The race of plants and the race of animals shrink under this great restrictive law. And the race of man cannot, by any efforts of reason, escape from it. Among plants and animals its effects are waste of seed, sickness, and premature death. Among mankind, misery and vice. . . . The ordeal of virtue is to resist all temptation to evil. (1798, 20)

In the Malthusian world, where the human conditions of existence are not significantly alterable, humanity must look elsewhere than to nature for its ethical ideal. Malthus could only resolve this dilemma with the aid of theodicy. If there was little moderation in nature, the burden of moral restraint was shifted to the individual. Morality was not the emulation of natural processes; rather, the moral individual must transcend natural circumstances, including instincts.

In Huxley's *Evolution and Ethics*, the physical conditions manifesting the "cosmic process" have the same destructive potential for the individual and for society. Malthusian necessity, however, despite Darwin's powerful generalizations of its applications, seemed only half the law for Huxley, as for many Victorians. Although governed by the realities of a physical world, society could exert a material influence over the conditions of existence so as to soften their impact. "The whole meaning of civilization is interference with this brute struggle," wrote Toynbee in his posthumously published Oxford lectures, *The Industrial Revolution*: "We intend to modify the violence of the fight, and to prevent the weak being trampled under foot" (1884, 59). Huxley likewise reaffirmed Malthus, only to call for human intervention: "Let us understand, once for all, that the ethical progress of society depends, not on imitating the cosmic process, still less in running away from it, but in combating it" (p. 83). Hence, in Huxley's concept of society, humans challenge the Malthusian landscape of conflict, not simply by individual restraint, as for Malthus, but by social and technological initiative.

Both Malthus and Huxley see the individual as a compound being, driven by instinct as well as by reason. For Malthus, the powerful organic cravings of the individual can be expected to oppose the general interests of society (1798, 163–64). Huxley likewise argues that the conflict between instinct and civil behavior remains an enduring challenge to civilization. But for Malthus, such disharmony furnishes a creative tension by which the individual is transformed in a teleology of self-transcendence: "Evil seems to be necessary to create exertion, and exertion seems evidently necessary to create mind" (1798, 204). The imperfections of nature are "admirably adapted to further the high purpose of the creation and to produce the greatest possible quantity of good" (ibid., 212). Malthus argues by theodicy that evil has purpose and is a necessity. In *Evolution and Ethics*, Huxley unceremoniously discards this teleological superstructure ordained by final end. He cannot understand why "among the endless possibilities open to omnipotence, that of sinless, happy existence among the rest, the actuality in which sin and misery abound should be that selected" (p. 72). The disharmonies of nature may drive human ingenuity and awaken the will to struggle against the cosmos, but there is no intelligence behind the reality. Evil is not the necessary source of greatest good, but only the origin of prodigious waste. England is not now, nor ever destined to be, the best of possible worlds.

Within the Malthusian conflict in the physical world of *Evolution and Ethics*, Huxley derived his rationale for ethics from the naturalistic categories of Hume. In his monograph *Hume* (1878), Huxley had traced his own agnosticism to Hume's skepticism, which demanded "the limitation of all knowledge of reality to the world of phenomena revealed to us by experience" (1893–94, 6:71). This pragmatic philosophy of experience had once obliged Huxley and his physiological colleague William Carpenter to dwell on Darwin's lack of physiological evidence for natural selection (1893–94, 2:16– 18, 72–76; Carpenter, 1889, 109; see also Hull 1973, 49–50; Bartholomew 1975; Ghiselin 1969; and di Gregorio 1984). And it appears intact in *Evolution and Ethics* in Huxley's conclusion that the facts of daily life confirmed the logical absurdity that humans and nature are antagonists. Huxley had always elevated experience over logic and theory. In *Hume* he had highlighted Hume's call for thinking individuals to "reject every system of ethics, however subtle or ingenious, which is not founded upon fact and observation (1893–94, 6:230; Hume 1777, 174–75). Hume's attack on systems is reiterated in Huxley's relentless attacks on utopian ethical theory in *Evolution and Ethics*. To Huxley, society is a built-up, makeshift, and therefore contractual affair, quite beyond the deliberate constructions of human method.

Ignoring the religious sanction for morality, Hume had naturalistically derived the moral sense from an innate sympathy, which in the social sphere took the utilitarian form of justice (Hume 1977, 180). "Crime or immorality," he wrote, "is no particular fact or relation, which can be the object of the understanding, but arises entirely from the sentiment of disapprobation, which, by the structure of human nature, we unavoidably feel on the apprehension of barbarity or treachery" (ibid., 292–93). Justice was a social concept, and hence artificial. As societies developed, Hume pointed out, mutual agreements would give advantages to those that functioned best as wholes (1893–94, 6:232). This notion of innate sympathy was adopted by Smith and Darwin as the likely origin of moral behavior, whose considerable utility in strengthening society would naturally lead humans to reinforce and codify sympathy in a system of

13

justice (Smith 1759, 87–89; Darwin 1871, 84–85; see also Ruse, 1986).

Sympathy, innately received, likewise furnishes the intuitive source of moral behavior in both essays of *Evolution and Ethics* (pp. 28–30, 79). In the "instinctive intuitions," Huxley found a unification of biology and philosophy. He incorporated Hume and Darwin in the argument that moral behavior was based on the intuitive responses of an evolved psychological faculty. "Cosmic evolution," he argued, "may teach us how the good and the evil tendencies of man may have come about; but, in itself, it is incompetent to furnish any better reason why what we call good is preferable to what we call evil than we had before" (p. 80; cf. Hume 1777, 214). And he went on to associate human notions of the good with the "evolution of aesthetic faculty," which by intuition—and not understanding—distinguished what is beautiful from what is ugly. As Darwin had done in his *Descent of Man*, Huxley cited Smith's *Theory of Moral Sentiments*, which made the innate human sympathy borrowed from Hume the coin of the ethical realm (p. 28; Darwin 1871, 81). Smith, Darwin, and Huxley all treated sympathy as an instinct, and hence, a source of pleasure (Smith 1759, 3, 7; Darwin 1871, 81).

To the two naturalistic principles of Malthus's natural inequality and Hume's socially redeeming sympathy, Huxley joined Hartley's doctrine of association to suggest a dynamic principle of psychological change. Huxley's philosophical views had always been distinguished by his interests as a physiologist in reflex action and instinct. He had published two textbooks on physiology, *Elementary Lessons in Physiology* (1866) and *The Crayfish: An Introduction to the Study of Zoology* (1879). His *Crayfish*, a classic of nineteenth-century physiology, was written at midcareer, roughly at the time he was writing his monograph *Hume* (1878) and its companion piece, "On Sensation and the Unity of Structure of Sensiferous Organs" (1879). In the latter article, he argued that sensations were modes of motion initiated from outside the body, that the brain was modified epithelium, and that "our sensations, our pleasures, our pains, and the relations of these, make up the sum total of the elements of positive, unquestionable knowledge" (1893–94, 6:318). This theme of the material basis of mind permeated Huxley's more sensationalist essays, such as "On the Physical Basis of Life" (1868) and "On the Hypothesis that Animals are Automata" (1874), in which he had argued that mental phenomena are known to us primarily, if not exclusively, by means of material phenomena.

Hartley's materialistic theory of individual moral growth by nervous association of experiences suggested a physiological basis for the evolution of ethical behavior in both individuals and society. In Hartley's theory of moral development, the social self emerged by degrees with the suppression of one's own natural cravings—which Huxley had identified with "original sin" (1892). In his *Observations on Man*, for example, Hartley had spoken of the "method of destroying the self, by perpetually substituting a less and purer self interest for a larger and grosser," in a process of associationist sensation building by degrees to morally more complex ideas (Hartley 1749, 2:282). The accumulated nervous experiences of the organism led to forms of more complex behavior, moral sense developing from simple sensations (Willey 1940, 137). A nineteenth-century physiologist like Huxley, contemplating evolutionary process, could interpret this process as adaptive.

Beginning with Hartley's view that the self was destroyed as the consequence of acculturation, Huxley went on in *Evolution and Ethics* to argue that the individual suppressed instinct by denying expression to an authentic part of the natural self. He characterized this process in terms of Smith's "ideal man within the breast"—the cultivated "artificial personality" of civilization (Smith 1759, 208; p. 30). In Hartley's idea of self-destruction, Huxley saw a principle of evolution of the artificial personality, in which benevolence is by degrees associated with perceived self-interest and becomes a source of the polity. Children, he noted, always come into the world entirely selfish, in the primitive manner of ancestral or savage humankind, but they gradually accept new social priorities: "every child born into the world will still bring with him the instinct of unlimited self-assertion. He will have to learn the lesson of self-restraint and renunciation" (p. 44).

In framing his naturalism in the context of the eighteenth-century thought of Malthus, Hume, and Hartley, Huxley attempted to reveal the historical depths of the intellectual and social tradition of which he considered himself a part. With Malthus's secularization of fallen nature in the so-called natural inequality, Hume's naturalization of grace in sympathy, and Hartley's physiological theory of individual moral development, Huxley built a formidable naturalistic framework for the emergence of human ethical behavior. He used eighteenth-century thought most systematically to displace classical Christian theological speculation about the ethical dilemma of humankind. This displacement, suggesting the historical failure of Christian ethical theology, fit the model of uniformitarian overtaking catastrophic causal explanation (See also Barton 1983).

Huxley's elaborate secularization by means of systematic intellectual displacement was so thorough that Lightman has been moved to argue, in a fascinating tu quoque retort, that Huxley, in replacing one set of middle-class religious orthodoxies with another set of scientific orthodoxies, was formulating a new "religion," complete with a "holy trinity" (Lightman 1987, 176). In this view, which does some reductive violence to the concept of religion, Huxley's naturalistic scheme merely supplants a theological theodicy with a secular one that argues for status quo. The effort, Lightman argues, is "to reconcile people to the existing social order by conceiving of society as an organism that should be allowed to develop on its own accord, since it is slowly progressing and growing due to the irresistible movement of natural laws" (1987, 117–18). This laissez-faire argument of irresistible progress, long associated with Spencer, was explicitly rejected by Huxley in 1871 and thereafter on a regular basis (1893–94, 1:268–70). We must also keep in mind that Huxley's systematic and frequent use of religion as a comparative frame of reference for his own naturalistic thought was a rhetorical device for ridicule that served to juxtapose, in elaborate mock-heroic fashion, the sacred and the secular.

Huxley and the Victorian Anthropological Perspective

While drawing on eighteenth-century thought to frame his own naturalistic world view, Huxley shared with many of his contemporaries the understanding that cultures were dynamic. An important Victorian ethnologist in his own right, Huxley viewed civilization not as a fixed state, as Malthus, Hume, and Hartley had, but as a process that continually transformed the conditions of the natural world. In asserting the naturalistic basis of cultural evolution, Huxley both intensified the polarities between the primitive and civilized human and bound the two together in an unprecedented physical and psychological intimacy.

Huxley's ideological origins derived partly from the thought of the bourgeois historian François Guizot, whom Huxley, in his retrospective essay "Agnosticism" (1889), rated one of the most important influences on his thinking (1893–94, 5:235). Agnosticism, Huxley argued, was a product of the intellectual progress Guizot identified with history itself. Guizot had argued that civilization was "properly a relative term" that implied "both a state of physical well-being and a state of superior intellectual and moral culture" (Guizot 1828, 1:18). Propelled by the innately motivated desire of humanity for "political liberty" and "free inquiry," civilization was also attended by certain necessary restrictions of individual freedom (ibid., 1:74, 265, 290).

This theme of progressive civilization and its attendant restrictions runs through Huxley's entire program of cultural criticism, reaching its summit in *Evolution and Ethics*. On the one hand, the argument of the improved "external" condition of humans conformed to the social rationale of his essays on the progress of science; on the other hand, the theme of the progress of the intellect fit Huxley's model of an increasingly rational and competent society, founded on naturalism and physical science. These concepts gave a social and intellectual legitimacy to science, locating the man of science within the texture of history and furnishing him with an essential role in contemporary culture (see Turner 1974, 10–13). Guizot's ideology of transition enabled Huxley, in such essays as his "Prologue to Controverted Questions" (1892), to view science as the logical extension of the humanist critical spirit (1893–94, 5:14, 41).

For Guizot, civilization was "an improved condition of man resulting from the establishment of social order in place of individual independence and lawlessness of the savage or barbarous life" (1828, 1:18). His antithesis of the civilized and savage life, identical in some respects to the dynamic views of his British counterparts Edward Gibbon and Henry Buckle, was easily transformed to a conflict between mind and nature. In their inductive histories, Gibbon and Buckle, by a massive coordination of facts derived from long-term conflicts, identified historical process with natural and social processes that operated over extensive periods. Gibbon's Rome succumbed to physical and social antagonisms, including "the injuries of time and nature" and "the use and abuse of materials," "which continued to operate in a period of more than a thousand years" (Gibbon 1901, 3:863). John Stuart Mill's brilliant essay, "Nature," had appeared in 1874, arguing that "nature impales man, breaks him as on the wheel" (Mill 1969, 375). Buckle argued, in his naturalistic *History of Civilization in England* (1857–61), that "man is affected by four classes of physical agents; namely climate, food, soil, and the general aspect of nature" (Hanham 1970, xv). Gibbon's "injuries of time and nature," Mill's antagonistic forces of nature, and Buckle's "general aspect of nature" were opposed to human constructs in precisely the way that in *Evolution and Ethics* Huxley's cosmic process, manifested socially in the forces of instinctual aggression and physically in those of nature, opposed human artifice (see also Stanley 1957).

The process historians' concepts of physical agency were reinforced in many ways by the emerging work of British ethnologists and anthropologists at midcentury, work in which Huxley himself had played a significant role. These movements, secure in the assumption of British cultural superiority, contained seeds of this assumption's destruction in the naturalistic view that humans and their cultures, whether contemporary British or ancient Hindu, were material constructs that yielded much to comparative analysis. Such were the insights brought through empire.

Huxley's earliest work as a morphologist fitted solidly into the emerging disciplines of comparative zoology—morphology, physiology, and embryology—rising from the work of Buffon, Cuvier, Von Baer, Richard Owen, and E. Geoffrey Saint-Hilaire. In spite of themselves, these individuals often projected humans solely as physical entities (Greene 1959a). In *Man's Place in Nature*, Huxley maintained the psychical unity of humans with lower animals, as well as the monogenist racial theory of the common origin of man, which led Darwin in his *Descent of Man* to cite Huxley's volume extensively in support his own evolutionist view of humans (Darwin 1871, 13–17, passim; cf. Di Gregorio 1984, 153–58).

In midcentury London, the emergence and amalgamation of the ethnological and anthropological societies established new approaches to the study of cultures (Burrow 1966; Stocking 1971). In contrast to the humanistic study of classical cultures on the basis of texts, the new cultural anthropology now empirically examined actual societies for insights into physical and prehistoric culture (Stocking 1987). Darwin, Alfred Wallace, and Huxley all helped to legitimize this trend in their voyage narratives and studies. Darwin's Fuegians in the *Journal of Researches* (1839) had become standards of savage behavior, to which Darwin would ultimately return in his *Descent of Man* (1871). Wallace had written extensively about native peoples in papers, as well as in his *The Malay Archipelago* (1869). Huxley, during his voyage on HMS *Rattlesnake*, (1846–1850), had filled numerous notebooks with ethnological observations and drawings of native peoples in New Guinea and Australia, which he turned to account in his Fullerian Lectures on ethnology at the Royal Institution in 1866–1867. In a key paper, "On the Methods and Results of Ethnology," he assessed the state of the field in 1865 as an ethnologist in the Darwinian cast who considered humans an evolved form and held to the monogenist theory of human origins (1893–94, 7:248–52).

The new approaches to historical and anthropological analysis acted synergistically to establish a powerful new Victorian critical perspective, in which cultures, despite assumptions of English superiority, were seen relativistically as human constructs. Combining empirical observation with the comparative method and growth analogy, the new anthropologists brought the human physical species increasingly into phase with civilization. "The thing that has been will be," wrote Edward Tylor, "and we are to study savages and old nations to learn the laws that under new circumstances are working for good or ill in our own development" (1871 1:158–59). John Lubbock held that the conditions and habits of savages resembled those of ancestors and helped to reveal how current customs and ideals were lodged in the mind "as fossils are imbedded in the soil" (1869, 1).

Efforts to join present to past naturalized the contents of the mind by associating them with the primitive, prehistoric past. Tylor wrote, in the uniformitarian language of Charles Lyell: "The savage state in some measure represents an early condition of mankind, out of which the higher culture has gradually been developed or evolved, by processes still in regular operation as of old, the result showing that, on the whole, progress has prevailed over relapse" (1871 1:32). For Tylor, Victorian culture could be linked in fundamental and revealing ways with primitive cultures: "The civilized mind still bears vestiges neither few nor slight, of a past condition from which savages represent the least and civilized man the greatest advance" (ibid., 1:68–69). Although this view assumed a hierarchy of social and cultural characters with Victorian civilization at the apex, it nevertheless made the psychological connection between all humans (see also Bowler 1986, 78–80). Civilization was a fluid state, manifested in different degrees and modes by all societies, which were now, in some sense, on a par. Civilization, always in flux, could either grow or atrophy.

These lines of nineteenth-century historical and anthropological thinking were incorporated by Huxley into *Evolution and Ethics*, with its naturalistic themes of the dynamic interaction of nature and culture, the physical status of the human animal, and the evolution of religion and morals. To derive the moral categories of Hume, given the destabilized physical nature of Malthus and Darwin, Huxley imagined a psychically divided individual who incorporated the innate aggressive impulses of the ancestor with the acquired social restraint of the cultured being. He used faculty psychology to incorporate two versions of humanity into the same mind.

Huxley's view of nature in *Evolution and Ethics* is darker, more pessimistic, than it had been in his early career. This shift, which can be traced to several political, personal, and intellectual factors, was by no means straightforward. In his essay, "A Liberal Education and Where to Find It" (1868), he had optimistically personified nature as the "calm strong angel" who was playing at chess with man and who would "rather lose than win" (1893–94, 3:82). Still, Nature opposes the individual. The prototype of his angel had been the fiend in the famous painting of the German artist Moritz Retzsch, mockingly playing at chess with man for his soul. Three years later, in his essay "Administrative Nihilism," Huxley rejected the nature-as-norm concept insofar as it implied laissez-faire principles of society. Human aggression, he held, was a natural component of human nature and required social amelioration. Huxley's views of human nature here were possibly influenced by Bagehot's *Physics and Politics* (1869), in which the natural human was little more than an abject, aggressive brute.

In "Science and Morals" (1886), Huxley returned to the argument that "the safety of morality" lay not in theological creeds, but rather "in a real and living belief in that fixed order of nature which sends social disorganization upon the track of immorality, as surely as it sends physical disease after physical trespass" (ibid., 9:146). If the implication here was to follow nature, his 1888 essay, "The Struggle for Existence in Human Society," abandoned this last vestige of teleological speculation. In this essay, Huxley firmly seized upon the Malthusian factor of overpopulation as the fundamental destabilizing force of society and history (ibid., 9:208). Nature from the perspective of the primitive individual was the personification of Istar, the bloodthirsty goddess, who "wants nothing but a fair field and free play for her darling, the strongest" (ibid., 9:207). Yet, from the perspective of modern science, nature was a process of material logic:

> If we desire to represent the course of nature in terms of human thought, and assume that it was intended to be that which it is, we must say that its governing principle is intellectual and not moral; that it is a materialized logical process, accompanied by pleasures and pains, the incidence of which, in the majority of cases has not the slightest reference to moral desert. (ibid., 9:202)

Huxley now removed the moral content he had once invested in natural process. As Hume had long before argued, nature could be considered the term for that which *is* but not a term for that which *ought* to be (Hume 1739, 469). In virtue of its uniformity, natural process was open to logical analysis and manipulation, but the collective forces known as nature were nonteleological, without moral purpose. Hence, the ethical law could not be identified with the physical law. Ethics, Huxley went on to argue in *Evolution and Ethics*, cannot be "applied natural history" (p. 74).

Various personal, political, and intellectual factors all help to explain these shifts in Huxley's views. His daughter Marian's death in late 1887, two months before he wrote "The Struggle for Existence," must have had some effect on Huxley's state of mind. Not only had this favorite, much-animated, daughter been happily married and possessed of considerable artistic talent, but her long suffering could hardly have had anything to do with individual moral desert. Huxley's essay also reflected contemporary social unrest. Mob rule in the West End of London after the riot of February 8, 1886, attended by looting and robbery, was followed by several days of panic among Londoners, fueled by rumors of wandering hordes of toughs and impending revolution (Jones 1971, 291–93). This anarchy was repeated in the fall of 1887, at the height of the depression, with the Trafalgar Square riot, not only impressing upon middle-class Londoners the implausibility, not to mention danger, of laissez-faire social policy, but also inspiring a great round of speculative programs. A direct reflection upon these events, Huxley's "Struggle for Existence" essay asserts not only that humans are subject to the same destructive impulses witnessed in the natural world, as they seek to fulfill fundamental needs, but also that society must be a collective effort to substitute a state of mutual peace for one of war (1893–94, 9:204). The destabilizing element Huxley identified in his essay was the Malthusian tendency toward overpopulation, which was based on a universal human instinct.

Huxley's view of the natural human as beast played an important polarizing role in the social construction of his state of art, by identifying a naturalistic source for the antisocial tendencies of the individual. This comparative thesis of innate aggressiveness had a certain autobiographical cogency, given Huxley's own confrontational personality and sense of youthful vulnerability (1900, 1:5). Victorian social theory was often compounded partly of autobiographical elements, not merely in the projection of one's own personality, but also because narrative was a formidable means of constructing possible chains of causality (Landau 1984; Beer 1983). "The simplest study of man's nature," Huxley assured the largely Christian audience of his *Man's Place in Nature* (1863), "reveals, at its foundations, all the selfish passions and fierce appetites of the merest quadruped" (1893–94, 7:155). These moral dualisms appeared in many forms familiar to Victorians. Bagehot, who had drawn on Huxley's notion of artificial reflex actions as a means by which humans "stored faculty and acquired virtue," had argued in *Physics and Politics* that among primitive humans "the strongest killed out the weakest, as they could" (1869, 6, 18). In the "Struggle for Existence" essay of 1888, Huxley considered the ethical individual to be exposed to the "deep-seated organic impulses which impel the natural man to follow his non-moral course"; unremitting human misery inevitably brought humans to the verge of a "chaos of savagery" (1893–94, 9:205, 214).

Huxley's natural human made his final appearance in *Evolution and Ethics* as the brute, a deterministic instrument of instinct, with little independent will or consciousness, a "superb animal" who owed his success in the savage state to "his exceptional physical organization; his cunning, his sociability, his curiosity, and his imitativeness; his ruthless and ferocious destructiveness when his anger is roused by opposition" (pp. 51–52). The Victorian natural man was, in fact, the personification of contemporary naturalism, an embodiment of the forces that Darwin, Lubbock, Bagehot, Mill and others had in their diverse ways imagined in the prehistoric landscape. This fascinating if threatening new version of humanity was usually some blend of personal experience and the contemporary "savage," who was seen as an entity of instinct and expediency. Darwin concluded in *The Descent of Man* that primal man was a "savage who delights to torture his enemies, offers up bloody sacrifices, practises infanticide without remorse, treats his wives like slaves, knows no decency, and is haunted by the grossest superstitions" (1871, 2:405). These strains all reflected the vast, dizzying fall, via Victorian ethnology, of natural man from his heights of romantic innocence to a new natural state of brutish "innocence."

In *Evolution and Ethics*, Huxley reconciled the natural with the ethical individual by merging them through the agency of society. Tylor had argued in 1871 that the civilized mind still bore vestiges of a primal past: "How direct and close the connexion may be between modern culture and the condition of the rudest savage" (1871, 159). Huxley argued that violent cosmic nature had been internalized as a psychical force, "born with us," the product of "millions of years of severe training." Hence, ethical nature may count on having a "powerful enemy as long as the world lasts" (p. 85).

This affiliation of primitive and modern mentalities lent considerable importance to human constructive efforts to maintain the social whole, for it removed all sense that culture was inevitable. Such a theme had been taken up by Huxley's protégé E. Ray Lankester in his provocative monograph *Degeneration: A Chapter in Darwinism* (1880), which evoked the biological phenomenon of devolution as a metaphor for what might happen to civilizations. Using Gibbon's Roman Empire as a case in point, Lankester warned that humanity could "degenerate" into a contented life of material enjoyment accompanied by ignorance and superstition. At the beginning of the "Prolegomena" Huxley argued that progress is relative (p. 4). If human culture could progress, it could also persist, unchanged, just as it could degenerate. There was neither necessity nor unified purpose to evolution. These themes, which introduced new biological metaphors of instability to cultural discourse, are sources of the pervasive feeling of impermanence that suffuses Huxley's *Evolution and Ethics*.

HUXLEY, DARWIN, AND THE ETHICS OF UTILITY

In *Evolution and Ethics*, Huxley challenged utilitarian moral theory and its related late nineteenth-century theories of eugenics. What seems to have concerned him most was the threat to ethical codes posed by the convergence of (1) classical utilitarian theories holding behavior to be motivated by self-interest and pleasure, and (2) evolutionary theory and its utilitarian cycles of struggle and self-aggrandizement. This convergence could—and did for some, like Spencer and Galton—justify the radical pursuit of self-interest as a legitimate social doctrine.

"Blessings are provided for offspring by due self-regard," argued Spencer in his *Principles of Ethics* (1897, 1:222). The phrase "ethics of evolution," Huxley countered, was ambiguous, because it suggested that a code of ethical behavior could be based on the view that "fittest" was "best." But "fit" had no moral authority, since what was fittest depended entirely on the conditions of existence (pp. 80–81). This uncoupling of the fit and the moral put Huxley in conflict not only with Spencer and Galton, but also, to a considerable extent, with his own early view that nature was just.

Victorians widely associated the adaptive operations of evolution with the principle of utility (Wallace 1900; Richards 1982, 48, 50; Bowler 1984, 144–45; Young 1985, 3, 261–62), although many, including Darwin, had trouble accepting a strict utilitarian rationale for moral behavior. The broad social utility of moral behavior was rarely at question, since the stable, self-restrained community founded on mutual regard promised collaborative advantages in adaptation. More problematic was the individual basis of utilitarian morality, which was associated with ideas of pleasure, self-interest, and expediency.

Pleasure, as a general animal experience, a biological universal, so to speak, had provided a powerful conceptual tool for viewing human morality as an extension of animal behavior. But how could the pleasure-pain dynamic account for the renunciation required of an ethical individual in pursuit of the ideal signalled by *ought*? This key Victorian moral dilemma was faced by Darwin, Huxley, and Spencer alike, with very different results.

The deeply naturalistic tendencies of eighteenth-century ethical speculation nicely converged with the biological thinking by which Victorians like Darwin, Spencer, and Huxley absorbed classical debates over ethics in a new behavioral terminology of instinct and habit (see Moore 1979, 156; Richards 1982). For many, the biological universals of pleasure and pain, when associated with repetitive actions, introduced principles of reinforcement and habit that justified increasingly complex social behaviors, whether in the long-term sense of the development of species or in the short-term sense of the education of the individual.

Two assumptions found throughout Darwin's *Descent of Man*, Huxley's *Evolution and Ethics*, and Spencer's *Principles of Ethics* set the behavior of humans and animals on a continuum out of which a principle of behavioral development could be extracted: the unity and variability of animal and human faculties. Darwin argued in specific terms that the mental faculties of higher animals and humans were identical: "Man and the higher animals, especially the primates, . . . have the same senses, intuitions, and sensations—similar passions, affections, and emotions, even the more complex ones" (1871 48–49). This linking of the mental faculties of humans and animals, together with the principle of variability, supported a naturalistic concept of cultural evolution consistent with the thought of anthropologists like Tylor and Lubbock, whom Darwin cited frequently.

In his *Descent*, Darwin derived moral behavior from a "complex" of mental and social factors grounded in "the social instincts [which] lead an animal to take pleasure in the society of its fellows, to feel a certain amount of sympathy for them, and to perform various services for them" (ibid., 72). Citing Hume and Smith, Darwin argued that sympathy had distinct adaptive value, for those communities which included the greatest number of sympathetic members, would flourish best and rear most offspring (ibid., 81–82, 85; see also Schweber 1977; Manier 1978). Yet, as Darwin also understood, these two principles embodied a potential conflict between the individual and the community (see Richards 1982). Behavior that was adaptive for the community as a whole was not necessarily beneficial or pleasurable for the individual; the reverse was true, as well. If individual instinct was the driving force of moral behavior, Darwin concluded, communal opinion, in directing that instinct, must often deflect it (1871, 72, 93, 98–99).

Darwin clearly saw the conflict between various human instincts, but he viewed this conflict as one that steadily diminished with experience and culture. This primeval conflict in humans was still part of civilized experience: "At the moment of action, man will no doubt be apt to follow the stronger impulse, and though this may occasionally prompt him to the noblest deeds, it will far more commonly lead him to gratify his own desires at the expense of other men" (ibid., 91). The processes by which behavioral priorities passed from the individual to the community thus became central to Darwin's theory of moral development:

> With increased experience and reason, man perceives the more remote consequences of his actions, and the self-regarding virtues, such as temperance, chastity, &c., which during early times are, as we have before seen, utterly disregarded, come to be highly esteemed or even held sacred. . . . Ultimately, a highly complex sentiment, having its first origin in

the social instincts, largely guided by the approbation of our fellow men, ruled by reason, self-interest, and in later times by deep religious feelings, confirmed by instruction and habit, all combined, constitute our moral sense or conscience. (ibid., 165–66)

This series of increasingly directive, institutionally reinforced moral behaviors suggests the gradual overtaking of instinct by consciousness. It further suggests the eventual conversion of individual priorities into those of the generalized whole or community. Darwin concludes, in optimistic, even utopian language, that the struggles in men "between higher and lower impulses" will with generations become less severe and "virtue will be triumphant." The "Golden Rule" is the natural outcome of social instinct, aided by "active intellectual powers and effects of habit" (ibid., 104, 106; see also Greene 1977).

Darwin's formula for the evolution of the moral sense or conscience in humans has a distinct strain of anthropomorphism—if not outright autobiography—that suggests the educational narrative of a Victorian youth raised in affluence. For Darwin, as for Huxley, the moral development of the individual—partly conceived in terms of one's own biography—recapitulates the psychological and cultural development of the race. In the infant, Darwin observed, "we may trace the perfect graduation from the mind of an utter idiot, lower than that of the lowest animal, to the mind of a Newton" (ibid., 106). To a considerable extent—with due regard for available facts—moral development seems to have been imagined by Darwin and his Victorian counterparts by reading personal experiences and values into the processes of nature so as to derive "naturalistically" their own culture and moral norms (see Landau 1984; Flew 1967; Beer 1983).

Despite his general agreement with utilitarian moral theory, Darwin explicitly rejected J. S. Mill's view in *Utilitarianism* that the moral sense was acquired and not innate (1871, 71). If moral behavior had evolved partly by the natural selection of instincts, one could answer the moral dilemma of *ought*—actions that often went against the individual's immediate interests—for the evolution of complex instincts through the natural selection of variations in simpler instincts would establish a nonconscious basis of moral behavior. Such a phenomenon could explain how more complex social instincts developed in "sterile worker-ants and bees, which leave no offspring to inherit the effects of experience and modified habits" (ibid., 38). The same phenomenon might explain a high standard of morality that gave slight or no advantage to the individual and his offspring over members of the same tribe, but gave advantages over other tribes. Hence, Darwin argued, "with strictly social animals, natural selection sometimes acts indirectly on the individual, through the preservation of variations which are beneficial only to the community" (ibid., 155). In the evolution of nonconscious human drives, the selection unit might be identified with the group (see also Ruse 1986, 218–19).

Huxley must have returned to the *Descent of Man* as he was writing *Evolution and Ethics*, for he took up the same issues he had considered in 1871, when defending Darwin's new volume against St. George Mivart and Alfred Wallace (Hull 1973; Wallace 1864). We find the same issues Darwin had taken up—the Malthusian pressures, the unity of human and animal faculties, the instinctual basis and utility of morality, behavioral mimicry, and sympathy. The faculty of moral sense, Huxley argued, is heightened through "mutual affection of parent and offspring, intensified by the long infancy of the human species," and socially generalized by the tendency of the individual "to reproduce in himself actions and feelings similar to, or correlated with, those of other men" (p. 28). The human individual, the most consummate of all mimics in the animal world, is also an emotional chameleon, who feels and needs sympathy and regards peer opinion as essential to his own well-being. Here, as in Darwin, we find strong biographical themes, blended with Huxley's distinctive behavioralist emphasis. Huxley gave more emphasis than most to nonconscious, irrational sources of human motivation. The themes of animal automation had long preoccupied him, especially the notion, pursued in his 1874 Belfast Address, "On the Hypotheses that Animals are Automata," that nonconscious, nonpurposive psychic components contributed substantially to human behavior (1893–94, 1:199–250).

In *Evolution and Ethics*, the themes of autonomic behavior resurface in Huxley's argument that the evolution of nonconscious behavior helps to explain the emergence of the moral sense. Criticizing Adam Smith for insisting on justifying social behavior on the basis of reason, Huxley retrieves Darwin's example of bee society to argue that complex social organization can develop through the evolution of nonconscious instinct. A thoughtful drone, he pointed out, "must needs profess himself an intuitive moralist of the purest water":

> He would point out, with perfect justice, that the devotion of the workers to a life of ceaseless toil for a mere subsistence wage, cannot be accounted for either by enlightened selfishness or by any other sort of utilitarian motives; since these bees begin to work, without experience or reflection, as they emerge from the cell in which they are hatched. (p. 25)

This "automatic mechanism" hammered out by natural selection, motivates behavior advantageous to the group, but not necessarily to the individual. Therefore, it cannot be justified solely by utility, contrary to the theories of Mill and Spencer. At base, Huxley speculates, human society was "as much a product of organic necessity as that of the bees" (p. 26).

For Huxley, as for Darwin, the development of moral behavior had two distinct components—instinctive behavior, reinforced by conscious intelligence, and acquired reflex. In his *Expression of Emotion in Animals*, Darwin drew on Huxley's distinctions between *natural* and *artificial* reflexes (Huxley 1866) to explain instinctual and habitual expressive behavior that was beyond volition. Darwin, however, took the Lamarckian view that acquired behaviors could be inherited, a view which Huxley and his former associate Lloyd Morgan rejected (Morgan 1896, 252). "Some actions," Darwin noted, "which were at first performed consciously, have become through habit and association converted into reflex actions, and are now so firmly fixed and inherited, that they are performed, even when not of the least use, as often as the same causes arise, which originally excited them in us through volition" (1872 39–40). Such reflex actions, he went on, "gained for one purpose might afterwards be modified independently of the will or habit, so as to serve for some distinct purpose" (ibid., 41). These Lamarckian tendencies were clear in Darwin's thought, although Darwin also argued that human expressive emotions could be based on nonconscious elements subject to selection.

Where Darwin accepted a Lamarckian mechanism for the migration of behavior toward the altruistic end of the moral spectrum, Huxley, with his faculty psychology and its material, autonomic components, divided the human mind against itself. This view provided Huxley the foundation for his thesis in *Evolution and Ethics* that innate determinants of behavior must coexist in the mind in considerable tension, not only with one another, but also with acquired behaviors. Such antagonism is the source of the dualism of *Evolution and Ethics* that troubled so many of Huxley's commentators (Stephen 1893; Dewey 1898).

Huxley accepted the notion of inherent human conflicts in motivation because certain contents of the mind were beyond the shaping influences of self or group. The human organism owed its success in the cosmos—not to mention in Victorian London—to faculties expressed in aggressive drive, quite as much as to those manifested in cooperative skills. Unlike Spencer and Darwin, who assumed that antisocial impulses would by degrees be converted into the substance of ethical behavior, Huxley argued that self-restraint and self-assertion were instincts with equal claims upon the normal, thriving individual and its society:

> Just as the self-assertion, necessary to the maintenance of society against the state of nature, will destroy that society if it is allowed free operation within, so the self-restraint, the essence of the ethical process, which is no less an essential condition of the existence of every polity, may, by excess, become ruinous to it. (p. 31)

The impulses of aggressive behavior remained essential sources of human drive—the will with which the individual combated the cosmos (cf. Helfand 1977, 176). Hence, virtue was not likely to "triumph," as Darwin had suggested. The Golden Rule, strictly observed, Huxley noted, subverts the interests of society because it refuses to punish law-breakers.

Huxley's anti-utilitarian sentiment is felt throughout the two essays of *Evolution and Ethics* in pervasive references to the pain and suffering of human existence. In Huxley's view, the pressures of social life require that individuals relinquish the prerogative of acting out their instinctive impulses. The very refinements among humans of emotion and intellect are attended by a proportional enlargement in the capacity for suffering. And, once again reflecting on themes of considerable personal import, he adds, "the divine faculty of imagination, while it created new heavens and new earths, provided [man] with the corresponding hells of futile regret for the past and morbid anxiety for the future" (p. 55).

This denial of the sufficiency of utility as a moral principle was directed at Herbert Spencer and his laissez-faire social policies. Spencer had linked ethical behavior, as had Mill, with the higher pleasures of self-sacrifice. To Spencer, the individual organism advanced by experience, which accumulated biologically through the generations. Although Spencer repudiated an absolute laissez-faire doctrine (1893 2:601), his social and ethical philosophy was rigorously anti-interventionist on the biological assumption that struggle was a means of adjusting and strengthening society. With his Lamarckian views that acquired behaviors could be transmitted, Spencer applied biological metaphors to society to derive a self-adjusting organism. He had increasingly less sympathy for the interventionist spirit of meliorative social policy (Burrow 1966, 206).

Spencer's theories, despite their elaborate tissue of doubtful biological metaphor, were often original and provocative, a fact that was not lost on Huxley. Spencer had grasped, for example, the idea that the preservation of biological essence through the generations was somehow tied up with the axioms of behavior for all organisms, including humans (1897 1:221). Thus, the superior claims of egoism were that they increased the chances of one's offspring's success. For Spencer, excessive altruism and self-denial led to such social evils as mental illness and the production of "relatively weak offspring" (ibid., 1:225–227). Huxley had once half-jokingly written to Spencer on Spencer's philosophy: "No objection except to the whole" (1900, 2:185).

The core of Spencer's ethical thesis is his identification of "survival of the fittest" with "utility" and hence, pleasure (1892, 1:352). Where the utilitarian ethics of Bentham and Mill had assumed the stability and universality of human nature, Spencer now saw it as variable and progressive (Greene 1959b, 432). Accepting the principle of greatest happiness as the basis of human motivation, Spencer argued that pleasure, variously secured, was the basis of the adaptive behavior by which humans had constructed a democratic society. By the progress achieved as individuals, constantly adjusted to new conditions of existence, egoistic and altruistic behaviors were found increasingly to converge. Hence, although Spencer held with Darwin and Huxley that altruistic and egoistic behaviors had evolved simultaneously, he argued further that "every species is continually purifying itself from the unduly egoistic individual, while there are being lost to it the unduly altruistic individuals" (1892, 1:234). By degrees, Spencer speculated, with only the vaguest Lamarckian mechanism in mind, egoism and altruism were to merge at higher stages of evolution in what he called the paradox that "the pursuit of the altruistic pleasure has become a higher order of egoistic pleasure" (ibid., 1:325).

Spencer's concept of behavior assumed a transformational dynamic that Huxley's did not. That is, new behaviors transformed the primal sources of motivation, and instincts were converted to the substance of ethical behavior. This view, central to John Dewey's interventionist thought and philosophy of education, led in Spencer to the conviction that conditions left to themselves were most likely to produce the human adaptive behaviors that would improve upon human nature itself.

A stark contrast emerges between Huxley's and Spencer's views in their respective concepts of humanity's future. Where Huxley stoically envisions society in the present as an imperfect but still effective instrument of ethical collaboration, attended by the pains of necessary renunciation, Spencer looks to the attainment of a future society in which the egos of individuals merge in the interests of the whole. For Spencer, the human is "undergoing transformation from a nature appropriate to his aboriginal wild life, to a nature appropriate to a settled, civilized life" (1897, 1:24). For Huxley, human nature is for all intents and purposes biologically fixed, given the slowness of evolutionary change; meaningful improvements must therefore be sought in the environment, transformed by human intervention into the state of art. Hence, for Huxley, laissez-faire policy *is* the essence of the struggle for existence; for Spencer, laissez-faire policy promotes conditions under which the human organism advances in perfection. Spencer's hope is in the change of human nature; Huxley's hope is in the change in human conditions.

Huxley thus concludes his "Prolegomena" with the observation that the "political animal" is "susceptible of a vast amount of improvement, by education, by instruction, and by the application of his intelligence to the adaptation of conditions of life to his higher needs." Yet the individual remains forever outside the best of possible worlds, "compelled to be perpetually on guard against the cosmic forces, whose ends are not his ends, without and within himself." And the prospect, Huxley concludes, "of attaining untroubled happiness, or of a state which can, even remotely, deserve the title of perfection, [is] as misleading an illusion as ever was dangled before the eyes of poor humanity" (p. 44). It is this final stoicism uttered by the companion of mortality, not some bourgeois political ambition on behalf of status quo, that ultimately moves Huxley to reject political radicalism.

Huxley's vision makes a revealing contrast with Spencer's curious alter ego, the human of the future:

> Bounding out of bed after an unbroken sleep, singing or whistling as he dresses, coming down with beaming face ready to laugh on the smallest provocation, the healthy man of high powers, conscious of past successes and by his energy, quickness, resource, made confident of the future, enters on the day's business not with repugnance but with gladness; and from hour to hour experiencing satisfactions from work effectually done, comes home with an abundant surplus of energy remaining for hours of relaxation. (1897, 1:220)

One is struck by the mechanical, formulaic quality of this futuristic figure, by the utter failure of realism, by the emotional sterility of Spencer's imagination. Such was the Victorian divergence in naturalistic concepts of the moral human individual.

In Spencer's naturalism, we see Lamarckian biology in service of the ideal human, yet another version of Quételet's average man of the future. It is a two-dimensional figure, the leader-on-the-poster, gazing amicably in the foreground of a flag. For Bagehot, it was the "national character" who, by some " 'chance predominance' made a model, and then invincible attraction, the necessity which rules all but the strongest men to imitate what is before their eyes, and to be what they are expected to be, molded men by that model" (1869, 26–27). It is classless man. In Huxley's non-Lamarckian naturalism, however, we see realistic, material man, struggling with his own imperfections against forces he only partly comprehends (1893–94, 9:ix). Huxley's is the "eternally tragic" figure, as seen in Oedipus or in Jude Fawley of Thomas Hardy's *Jude the Obscure*.

ART, ARTIFICE, AND THE ARTIFICIAL WORLD

Huxley's view of human nature as virtually fixed shifted the burden of progress to the transformation of environment. This transformation of the human conditions of existence, demonstrably made possible by artifacts, was one of the striking instances in which Victorian evolutionary and anthropological theory converged. The biological role played by technological intervention elevated human artifice to the status of a elemental force against what to Malthus had seemed a cosmic background of irresistible natural process. These newly apparent powers of transformation inspired an important line of Victorian social speculation concerning the manner in which humans routinely helped to shape their own destinies. Huxley not only participated in this discussion, but his *Evolution and Ethics* may be seen as its Victorian culmination, finding in the progress of human material construction one of the great social hopes of the future.

The role of human artifice was greatly magnified in Huxley's view by the increasing technological prowess of the industrial being. The argument of "Evolution and Ethics" is nested in the fable of Jack and the beanstalk, which Huxley made over into an allegory of the cyclical ebb and flow of civilizations. Out of a primary world of forces and matter rises a superstructure, a "world composed of the same elements as that below, but yet strangely new" (p. 46). This artificial world is none other than English industrial civilization, which Huxley in his famous essay, "The Progress of Science, 1837–1887," had presented as the "new nature, begotten by science upon fact." In Huxley's robust industrial optimism, we see a distillation of the Baconian transformation of the world, with one important difference. Progress, perpetually at odds with fundamental physical and instinctual forces, is not inevitable, nor is it toward anything of remotely millenial proportions. Thus, it is nonteleological. Victorian civilization is not the best of possible worlds, and its existence, given physical reality, must eventually cease.

The antithesis Huxley posed between society and the cosmic background made a paradox of humanity that troubled many commentators, including Spencer and Dewey. How could humans, the products of the cosmos, revolt against the cosmic force? Were they not, as Mill observed in his essay "Nature," physical beings and necessarily extensions of nature? (Mill 1969). Huxley's Promethean call for the revolt against nature must have seemed to many an expression of hubris—the unthinking arrogance of human technological intervention.

The notion of humans as independent, transforming agents in nature was forcefully spelled out in the first chapter of Darwin's *Origin of Species*, "Variation Under Domestication," with which Huxley, Wallace, and many other Victorians were familiar. Huxley's references throughout the "Prolegomena" to the "horticultural process" suggest that he had reread Darwin's chapter. Humans, the subjects of selection, had also become creative forces in nature. As artificers of life forms, they could adapt the accidents of nature to their own ends (Darwin 1859, 30; see also Ritvo 1987). Darwin's use in the *Origin* of artificial selection as the conceptual template for natural selection is well known (Young 1985; Ruse 1979; Kohn 1980). Alfred Wallace and, more recently, Robert Young have drawn attention to certain fallacies implicit in Darwin's active metaphors for nature as an agent of "selection" (Wallace 1864; Young 1985). We can see in Darwin's thinking at the very beginning of the *Origin* an incipient dualism in which the human intervention through horticultural art becomes the allegory of natural process. Human designs guided by human purposes provide the model for, and hence systematically shift, the nonteleological processes of nature. An anthropomorphic analogy presides over the stringent naturalism of Darwin's *Origin of Species* (Beer 1983).

The theme of human transformation in the first chapter of the *Origin* thus contrasts with the natural world in the remainder of the volume in ways that shed light on Huxley's dualistic argument in the two essays of *Evolution and Ethics.* In Darwin's domestic world of the English countryside, husbandry bends nature to human ends, whereas in the natural world beyond, environment is governed absolutely by the laissez-faire dynamics of struggle for existence. It is the very contrast Huxley made between the state of art and the state of nature, governed respectively by the horticultural and the cosmic processes. Hence, while Darwin's intentions in the *Origin* were not even remotely Huxley's in *Evolution and Ethics*, they both promoted the common theme of the power of artificial process to reconstruct material reality.

The power of artificial processes to supplant natural processes was a theme that many of Darwin's midcentury contemporaries explored as nature became increasingly difficult to accept as a social and cultural norm. For Alfred Wallace, who had trouble deriving human culture solely on the principle of natural selection, the dualism implicit in "natural selection" suggested a principle of cultural development. Unlike brutes, whose harmony with their surroundings was perpetually destabilized by new conditions of existence and variability in generations, humans fashioned tools to mediate their fit with the natural world.

> [Man] is, indeed, a being apart, since he is not influenced by the great laws which irresistibly modify all other organic beings. Nay more: this victory which he has gained for himself gives him a directing influence over other existences. Man has not only escaped "natural selection" himself, but he actually is able to take away some of that power from nature which, before his appearance, she universally exercised. We can anticipate the time when the earth will produce only cultivated plants and domestic animals; when man's selection shall have supplanted "natural selection"; and when the ocean will be the only domain in which that power can be exerted, which for countless cycles of ages ruled supreme over all the earth. (1864, p. 168)

Human directive will, made possible by conscious mind, could substitute human purpose for natural process and allow man with an unchanged body to alter the conditions of existence with increasing effectiveness, an argument Darwin followed in his *Descent* (1871,

158–59). Thus, we see the makings of a new teleology, by which material reality is permanently molded by human artifice (Turner 1974, 76–78).

Such transformational optimism promised a kind of utopian control that would redeem the natural world. Marrying the dynamic of evolutionary progress with the Baconian theme of the advancing human control of nature through new knowledge, many Victorian intellectuals, the young Huxley included, were inclined to read into the human condition a necessarily higher destiny. This ideological construction also correlated with the Marxist view of the conquest of nature through human labor amplified by technique (Engels 1927; Williams 1980). Indeed, it furnished an array of social theorists with the notion that humans, restored to their natural rights within less artificial social structures, and supported by material progress, could attain a civilization of near-millenial proportions.

Yet Victorian transformist ideology also delivered a hidden problem of self-restraint that was not seen very clearly by Wallace or his scientific and political counterparts. The problem, as the American philologist and diplomat George Perkins Marsh showed in his *Man and Nature* (1864), was that human intervention was not only more powerful than either Darwin or Wallace had imagined; it was also pathological. Human action upon nature, Marsh suggested, differs in essential character from animal action because "it is guided by a self-conscious and intelligent will aiming as often at secondary and remote as at immediate objects" (1864, 41). The human tenacity of purpose had a troubling potency: "There are parts of Asia Minor, of Northern Africa, of Greece, and even of Alpine Europe," Marsh noted, "where the operation of causes set in action by man has brought the face of the earth to a desolation almost as complete as that of the moon . . ." (ibid., 42). Marsh documented the creation of deserts, the extinction of animal populations, and vast alterations of geological structures, impressing Lyell with his evidence for the magnitude of the human capacity to alter geological nature. But Marsh could not explain the pathology of the human war against its environment.

This awareness that human powers of transformation were steadily advancing against nature moved the romantic humanist John Ruskin to near-hysteria in his attacks against profane utilitarian science (Sawyer 1985, 217–246). But transformist ideology continued to gain ground in the 1860s and 1870s among those who sought no normative referent in nature. In Darwin's thought and in the industrial revolution, both human and material nature had been seen to offer vast potential for transformation. W. K. Clifford, for example, wrote in his remarkable 1877 essay, "Cosmic Emotion": "Human nature is fluent, it is constantly, though slowly changing, and the universe of human action is changing also" (1877, 420). These instabilities offered an advantage to human enterprise, for if nature was uniform, collective human effort could aspire to control. "How far away is the doctrine of uniformity from fatalism!" Clifford wrote. "It begins directly to remind us that men suffer from preventable evils, that the people perisheth for lack of knowledge" (ibid., 417). To Clifford, evolution supplied the key to ethics: "My actions are to be regarded as good or bad according as they tend to improve me as an organism, to make me move further away from those intermediate forms through which my race has passed, or to make me retrace those upward steps and go down again." Drawing on Spencer's concept of the social organism, Clifford declared that human collective action—or "Band Work"—enabled the social whole to " 'originate events independently of foreign determining causes,' or to act with freedom" (ibid., 423). This notion of environmental plasticity under the collaborative stress of human intervention reemerged in the work of John Dewey (1920) and Alfred North Whitehead (1925). The Kingdom of Man was at hand.

Huxley's *Evolution and Ethics* was the product of the same tradition of transformational speculation that had followed in the wake of Darwin's *Origin of Species*. But by the 1890s, Huxley was reflecting back across an economic and social landscape in considerable disarray, as the result of the severe depression in England. The idea that social amelioration was to emerge in any automatic manner by the maintenance of laissez-faire social policies was no longer acceptable. Rejecting religion as a human effort to manipulate, by theodicy, the *concept* of reality rather than the reality itself, Huxley looked once more to science, which offered a means of neutralizing the mechanisms of natural selection. He accordingly rejected the theodicy—the assumption that this is the best of possible worlds—and challenged the cosmic process.

Huxley argued, as Wallace, Clifford, and Mill had, that modern humans construct an artificial world. Intervention and transformation were the great objects of civilization. By artifice, humans modified, albeit temporarily, the forces of nature and helped in the making of their own destiny. "The history of civilization details the steps by which men have succeeded in building up an artificial world within the cosmos. . . . In virtue of his intelligence, the dwarf bends the Titan to his will" (pp. 83–84). This argument drew on Huxley's 1887 essay, "The Progress of Science," in which he spoke of "a new Nature, the existence of which is dependent upon men's efforts, which is subservient to their wants, and which would disappear if man's shaping and guiding hand were withdrawn" (1893–94, 1:51). He identified the transformation as the "new birth of time"—the divergence of civil from natural history.

Huxley had difficulty, as did Lyell and Wallace, accepting the gradualist view of cultural development promoted by Tylor, Lubbock, and Darwin. For Huxley, this was because of the brevity of civil history, given the great cultural differences between the primitive being of his *Man's Place in Nature* (1863) and the modern intellect that worked the miracles of his essay, "The Progress of Science." Moreover, there was little evidence in historical time that the human mind was evolving (p. 77–78). In his essay "Administrative Nihilism" (1871), Huxley had thus argued a theory of cultural saltation, suggesting that some primordial "aggregation," a term he borrowed from Walter Bagehot and Henry Maine, might precipitate a condition or state of cooperative culture (1893–94, 1:274–75; Bagehot 1869, 17).

Huxley's state of art in "Evolution and Ethics" is a psychological construct, an existential condition engendered by a shared mentality, a social consciousness devoted to the persistence of humans both as material and as ethical beings. As Morgan pointed out, Huxley recognized in culture the emergence of a psychological principle of "mental evolution," which enabled conscious choice to displace natural selection as the primary agent of change (Morgan 1896, 334–36). Ethics, like art, worked in "harmonious contrast" with the forces of nature; both were governed by an innate aesthetic sense of the good or beautiful; both were secured at the expense of considerable self-restraint (pp. 31, 35, 80). Interventionist in spirit, these products of the human consciousness sought to manipulate the forces of the cosmos and subordinate them to human purpose. What is neutral in nature—for example, the innate self-centeredness of the biological individual—becomes culpable in the state of art. Neutral physical force and circumstance become tragic within the consciousness of art. Oedipus cannot exist outside the state of art, nor, for that matter, can Thomas Hardy's tragic figures. The ritual of human suffering reveals the very norm of nature, the cosmic process intruding upon the human condition (p. 59).

Thus Huxley, eschewing the unthinking anthropomorphism of religions, consciously reconstructs an anthropomorphic state of art, where the human becomes the moral center of existence. This is the object of all artifice. Huxley's psychological theory of culture as a mental state was consistent with contemporary Victorian speculation. Walter Bagehot, for example, had argued that humans exercised an increasing mental control over their physical limitations. This idea led to Bagehot's notion of the "age of discussion," in which static civilizations based on set customs were supplanted by progressive civilizations that placed priority on the exercise of intelligence—presumably in English parliamentary democracy (1869, 119). Similar but much more psychologically developed ideas were explored in Lloyd Morgan's theories of mental evolution. Morgan argued that ideas evolved not by selection but by the elimination of incongruity with other similar ideas in a cultural environment, which was wholly a psychic construct (1891, 485–94). D. G. Ritchie, in his *Darwinism and Politics*, explicitly extended themes of struggle he had found in Huxley's essay, "The Struggle for Existence in Human Society," to the domain of institutions. To Ritchie, institutions were artificial constructs that behaved as instruments of Lamarckian evolution, shifting the burden of transmission from biology to social artifices. "Natural selection operates in the highest types of human society as well as in the rest of the organic realm; but it passes into a higher form of itself, in which the conflict of ideas and institutions takes the place of the struggle for existence between individuals and races" (1891, 106). By such means, Ritchie argued, the well-being of society as an ethical end was substituted for the individualist conception of a balance of pleasures and pains (ibid., 106).

Huxley's psychological dualism accommodated a naturalistic, more complexly monistic concept of the civilized individual, who now emerged as both egoistically aggressive and altruistically self-less, irrational and reasoned. The social individual lived in considerable psychological tension, just beyond the sway of the instinctual self. One might thus reconcile the natural history of Darwin with the civil history of Gibbon.

Such psychological realism also revealed a pathological dimension of culture that Huxley was well aware of. The pervasive references throughout the two essays of *Evolution and Ethics* to the pains of renunciation were analogous to a feeling of illness, which Huxley's own long mental suffering, now augmented by the onset of heart failure and the loss of his daughter Marian, must have made palpable. Huxley had been attracted in his essay "Administrative Nihilism" by Immanuel Kant's theory in *Idea for a Universal History* (1784) that art was the product of the "unsocial sociability" of humankind. Huxley quoted at length from Kant's essay the argument that society "originated by a sort of pathological compulsion, [which became] metamorphosed into a moral unity." Kant had concluded that "all culture and art which adorn humanity, the most refined social order, are produced by that unsociability which is compelled by its own existence to discipline itself, and so by enforced art to bring the seeds implanted by nature into full flower" (1893–94, 1:277). In "Evolution and Ethics," Huxley modernized this argument: the civilized being necessarily denied expression to instincts that in a state of nature were fundamentally adaptive. In surpressing the "qualities he shares with the ape and the tiger," the divided individual had taken on the burden of pain, which reached its summit in quality and intensity among members of the organized polity (p. 51). Humans were thus compelled to live in the state of art at the expense of their own instinctual lives, living paradoxes, unable to reach under natural conditions the full potential of their powers.

The concept of human artifice as adaptive neither originated nor ended with Huxley. Yet Huxley transformed it into a dualistic principle familiar in twentieth-century organic theories of culture promulgated by individuals like Sigmund Freud and Konrad Lorenz. Freud, whose *Civilization and Its Discontents* took the theme of the irremediable antagonism between the demands of instinct and the restrictions of civilization, concluded his work in 1931 by observing:

> The fateful question for the human species seems to me to be whether and to what extent their cultural development will succeed in mastering the disturbance of their communal life by the human instinct of aggression and self-destruction. . . . Men have gained control over the forces of nature to such an extent that with their help they would have no difficulty in exterminating one another to the last man. They know this, and hence comes a large part of their current unrest, their unhappiness and their mood of anxiety. And now it is to be expected that the other of the two "Heavenly Powers," eternal Eros, will make an effort to assert himself in the struggle with his equally immortal adversary. But who can foresee with what success and with what result? (1931, 92)

In *Evolution and Ethics*, Huxley brought the logic of his generation to the verge of the considerable dilemma, also identified in the work of George Perkins Marsh, of humans at war with the cosmos, instinct joining with artifice against their fellows and environment.

The Reception of "Evolution and Ethics"

To many former associates accustomed to Huxley's robust team spirit in controversies with orthodoxy, "Evolution and Ethics" seemed an abandonment of the old "conspiracy" to translate questions into the monistic terms of scientific naturalism while invoking agnostic brackets for the larger ontological questions. Henry Drummond, the Scottish naturalist and evangelical writer, noted the widespread surprise at Huxley's dualistic drift: "For, by an astonishing *tour de force*—the last, as his former associates in the evolutionary ranks have not failed to remind him, which might have been expected of him—[Huxley] ejects himself from the world order, and washes his hands of it in the name of Ethical Man" (1894, 22). Others, like Lloyd Morgan, the comparative psychologist, found in

Huxley's views a crucial new distinction between human conscious choice and natural selection as agents of change (1896, 336).

Leslie Stephen, St. George Mivart, Herbert Spencer, John Dewey, Karl Pearson, and Petr Kropotkin, as diverse an assembly of late Victorian philosophical and social speculators as one could summon, all agreed that inconsistency had overtaken Huxley's thinking. With the exception of Mivart, the Roman Catholic evolutionist, they flatly rejected Huxley's psychological dualism, which set the instinctual content of the primitive mind against its transformed version in the social self and made humans the antagonists of nature. Psychological dualism, which Huxley had derived from nineteenth-century physiology and ethnology, recognized a complexity in the mind that others like Sigmund Freud would ultimately transform into a rich source of oppositions and submerged complexes. This idea of self-opposition naturalized the irrational and shattered one of the great dreams of Victorian social speculators, including Huxley himself: to unify the laws of nature and society under the principles of reason and utility.

Objections voiced both by Leslie Stephen and St. George Mivart in reviews of the Romanes Lecture demonstrated how thoroughly Huxley had confounded the proponents of liberalism and orthodoxy. Stephen held to the utilitarian line of his own *Science of Ethics* (1882), arguing that moral behavior was a product of evolution. He rightly noted that Huxley's metaphorical language anthropomorphically invested natural process with negative qualities—"ferocity," "evil," and "titan"—that infused nature with an intentionality it could not, in fact, support. Huxley's argument that the social individual was at odds with the natural individual was "awkward," Stephen suggested politely, for Huxley had introduced a developmental gap between the cosmic and horticultural processes that had no logical explanation. To Stephen, the social utility of virtue was identical with its evolutionary utility, making virtue the logical product of self-evident natural forces (1893, 164). If suffering, as Huxley had argued, was indeed a perennial companion of the higher organism, that pain was also a source of well-being. Stephen thus embraced theodicy: "the sheep may be, on the whole, the better for the wolf." (1893, 160).

As Stephen thus clung to his utilitarianism, so the Catholic evo-

43

lutionist Mivart rejoiced in Huxley's restoration of the gap between brutes and humans. Mivart, once the rising Huxley's anatomical assistant in the preparation of *Man's Place in Nature* (1863), had burned his professional bridges in a review of Darwin's *Descent of Man* that challenged Darwin's theory of the unity of human and animal faculties (Hull 1973, 372–75, 414; James R. Moore 1979, 120–21). Mivart had insisted in *The Genesis of Species* (1871) on a qualitative gap between the instinctive social behavior of animals and the conscious moral behavior of humans. Huxley had patronizingly dismissed Mivart's theistic morality and argued that both animal and human psychology reflected a utilitarian ethical strategy (1893–94, 2:170–71). But six years later, Huxley argued against strict utility that moral sense, like aesthetic sense, was intuitive (Huxley et al. 1877, 536–37).

In his 1893 review of Huxley's Romanes Lecture, "Evolution in Professor Huxley," Mivart noted with irony that Huxley's thought was itself "evolving." For Huxley was now challenging utilitarian ethics and insisting on an unexplained gap between natural process and human conscious will, between the cosmic and ethical process, between nature and man. Mivart, as he had done for twenty years, thrust God into this gap: "So great, indeed, is the contrast and distance between man and the world of irrational nature, that it suggests now, as it suggested of old, . . . a mode of being which is raised above all human nature, as man himself is raised above all infrahuman nature" (1893, 330). This was classical theistic argument. Merely beneficial social behavior, Mivart held in contrast to Stephen, could not be equated with conscious virtue. Rather, a new principle must come into play. Others thought similarly. The philosopher William Courtney argued approvingly that Huxley had left break enough between nature and humans for a spiritualist interpretation (Courtney 1895, 320–22). And an anonymous reviewer in *The Atheneum* marveled at Huxley's dexterity in promulgating a doctrine that made no mention of Christianity, yet "in its essential character was an approximation to the Pauline dogma of nature and grace" ("Evolution" 1893a, 119; see also Lightman 1987).

The shifts in Huxley's thinking detected by Stephen and Mivart were especially bitter to the embattled Herbert Spencer, whose Lamarckian theories of social progress came under increasing attack in the 1880s and 1890s by liberals, socialists, and physiologists alike. To Spencer, Huxley was deceitfully abandoning a tacit alliance that had served both men for a generation. "Evolution and Ethics," he reasoned in characteristic a priori fashion, could hardly be considered an attack against himself, since he and Huxley held identical views, and he had preceeded Huxley as a philosopher. It was unlikely that Huxley intended to teach him his own theories (1893, 184). Huxley, the same old utilitarian monist he had associated with for more than forty years, was evidently attacking the extreme anarchists. The footnotes of Huxley's published version of *Evolution and Ethics*, Spencer insisted, belied the insistent dualism of the main text. Huxley had stated in note 20 that "strictly speaking, social life, and the ethical process in virtue of which it advances towards perfection, are part and parcel of the general process of evolution, . . . just as the 'governor' in a steam engine is part of the mechanism of the engine" (Spencer 1893, 193–94; p. 114–15). Huxley, Spencer insisted, held views so similar to his own that the final paragraph of "Evolution and Ethics" expressed the very sentiment of the final paragraph of his own two-volume *Principles of Ethics*, which, Spencer noted pregnantly, Huxley had had in his hands some two weeks before he delivered the Romanes Lecture (1893, 194).

The long and fruitful association of Huxley and Spencer was too widely known for anyone to be surprised by Spencer's claim. As late as 1878 Huxley had written of his fellow X Club member for the ninth edition of *Encyclopedia Britannica*: "The profound and vigorous writings of Mr. Spencer embody the spirit of Descartes in the knowledge of our own day, and may be regarded as the 'Principes de la Philosophie' of the nineteenth century" (1893–94, 2:213). This was hardly the language of an antagonist, especially in view of the profound regard Huxley had for Descartes as the prototypical truth seeker (ibid., 5:166–98).

However, mutually supportive scientific naturalists as they were from a distance, Spencer's a priori method had always been at odds with Huxley's own stringent empiricism. For thirty years, their opinions had steadily widened over issues long identified with Spencer, which Huxley attacked in *Evolution and Ethics*: the law of progress, the perfectibility of man, laissez-faire social policy, radical individualism, and utilitarian ethics. In such seminal articles as "Progress: Its Law and Cause" (1857) and "The Social Organism" (1860), Spencer had made evolution and human social development parts of the same necessity, on the a priori assumptions that every force had multiple effects and therefore contributed to a growing heterogeneity that was the essence of progress, whether organic or social (Spencer 1904, vol. 1; Bowler, 1984).

In contradiction of the laissez-faire individualism such views inspired, Huxley promulgated theories of persistence, waged a vast campaign for public education and for improving natural knowledge, rejected the analogy of state and organism, attacked radical individualism, and finally discredited nature as an ethical model (see also Desmond 1982, 97–101). Indeed, Huxley's two essays, with their emphasis on environmental rather than organic change, were fundamentally at odds with the Lamarckian dynamic of Spencer's social views. Even Spencer's ardent Chinese disciple Yen Fu considered Huxley the antagonist and translated *Evolution and Ethics* into classical Chinese so as to provide a foil to the master's philosophy (Schwartz 1964).

What most troubled Spencer and other contemporaries about *Evolution and Ethics* was Huxley's final—and, really, quite characteristic—agnostic refusal to rationalize away the apparent gap between civilization and nature, ethics and instinct. Where Victorians had routinely invoked theodicies or teleological speculations to explain, justify, or diminish the discontinuity, Huxley agnostically abandoned the effort to provide it with a specific logic or faith, a point first noted by Lloyd Morgan (1923, 206). "If the conclusion that the [cosmic and horticultural processes] are antagonistic is logically absurd," Huxley had written, "I am sorry for logic, because, as we have seen, the fact is so" (p. 12). He thus accepted the concept of the irrational that others denied. Spencer had found in the gap teleological evidence for the "making of man," Darwin had normalized it with the principle of unity of mental faculties, Stephen had justified it in a theodicy, Mivart had cited it as evidence of spiritual intervention. But Huxley denied it an explanation.

In a letter of March, 1894, he observed:

> There are two very different questions which people fail to discriminate. One is whether evolution accounts for morality,

the other whether the principle of evolution in general can be adopted as an ethical principle.

The first, of course, I advocate, and have constantly insisted upon. The second I deny, and reject all so-called evolutional ethics based upon it. (1900, 2:360)

In his refusal to fill in the gap between the realms of biological determinism and human purpose, Huxley denied nineteenth-century theories of biological process the role of furnishing a social praxis (Himmelfarb 1968, 332; 1986, 79). But he did not suggest that there were no connections between the two realms, for instinct and physical nature were integral to the dynamic of which modern humanity was an inescapable part. The two essays of *Evolution and Ethics*, although based on Malthusian principles, sought to deny these primary roles in human society.

Huxley's social meliorism drew criticism from the nascent British eugenics movement. Leslie Stephen, oddly enough, was the first to raise eugenic objections to Huxley's call in "Evolution and Ethics" for the equipping of as many as possible to survive. Citing Karl Pearson on the struggle for existence among races of men, Stephen argued in his review of "Evolution and Ethics," "We give inferior races a chance of taking whatever place they are fit for, and try to supplant them with the least possible severity if they are unfit for any place" (1893, 170). In what appears to be an impressionistic blend of Spencer, Bagehot, Quételet, and Galton, Stephen spoke confidently, if theoretically, of developing the "best stock" of the race in a future "average man" who would combine great intellectual power with physical vigor: "We are engaged in working out a gigantic problem: What is best, in the sense of the most efficient, type of human being?" (ibid., 168; see also Quételet 1835, 96–97).

Huxley's anti-eugenicist "Prolegomena," although partly a response to Stephen's review of "Evolution and Ethics," broadly criticized the growing eugenics movement. Galton's *Hereditary Genius*, which appeared in its second edition in 1892, classified humans by qualities that Galton argued were transmitted among human "breeding groups" through the generations. Denying a role to nurture and giving all to nature in the physical and mental constitution of the individual, Galton raised hereditary transmission to a social absolute (Kevles 1985). He sought to formalize and "humanize" the "survival of the fittest" in a deliberated public policy of hereditary transmission. An oligarchy of breeding experts might manipulate—ever so slightly—conditions of human existence, so as to vary the birth and death rates of favored groups and thus improve the physical and mental qualities of the average human (Galton 1892, 27, 41). The purifying role of Spencer's laissez-faireism was thus supplanted by Galton's artificial selection.

Huxley concluded that the social interventions of the eugenicists, despite their Platonic concerns for the ideal individual, failed to imagine the brutal consequences for the real individual. The selection of humans by other humans, he argued, would destroy the bonds of sympathy that held society together, without which "there is no conscience, nor any restraint on the conduct of men" (pp. 36–37). Huxley's own family history of mental instability would not have entitled him to any favors in a Galtonian eugenicist state, as Huxley was well aware (Paradis 1978). "One must be very 'fit,' indeed," he noted, not to know of an occasion or two "when it would have been only too easy to qualify for a place among the 'unfit' " (p. 39). Huxley's anti-eugenicist stance was challenged by Karl Pearson in an address delivered at the Oxford University Junior Science Club nearly fourteen years later to the day, on May 17, 1907. The conflict between the ethical and cosmic processes, Pearson argued, was in reality a conflict between "human sympathy" and "racial purification." Human sympathy, he noted without a hint of irony, was overwhelming English society: "One factor—absolutely needful for race survival—sympathy, has been developed in such an exaggerated form that we are in danger, by suspending selection, of lessening the effect of those other factors which automatically purge the state of the degenerates in body and mind" (1907, 25). Huxley's state of art, by suspending selection and meliorating the physical conditions of existence, was to Pearson the certain source of physical, mental, and moral degeneration.

John Dewey, another avid Huxley-watcher, took the themes of environmental transformation and social intervention by education not as liabilities but as central arguments in a new philosophical scheme. The antagonism that Huxley saw between humans and their environment was a dualism that could be recast both as a dialectic between fact and ideal and as a source of social drive. Humans progressed, the young Dewey agreed, by reconstructing a part of the natural environment, not so much in rebellion against nature as "to relate a part of nature more intimately to the environment as a whole" (Dewey 1898, 326). Themes of transformation, both material and social, permeated Dewey's later philosophy of reconstruction. Nature was "something to be modified, to be intentionally controlled. It is material to act upon, so as to transform it into new objects which answer to our needs" (1929, 80–81; see also 1920, 116). But Dewey rejected what he saw as Huxley's dualistic view that brute self-assertion remained a determinant of human behavior. Rather, "the natural process, the so-called inherited animal instincts and promptings, are not only the stimuli, but also the materials, of moral conduct" (1898, 332–33). Like Spencer, Dewey believed that the animal impulses were not just deflected but actually eliminated by transformation. Dewey substituted for Huxley's gap between natural and ethical humans his own monistic behavioral gap between "habit and aim," arguing that the social priority was to transform the former to the latter (ibid., 335).

Demonstrating the formidable plasticity of biosocial theory itself, Petr Kropotkin, geographer, anarchist, and social theoretician, launched an ambitious frontal attack on Huxley's assumptions by reconstructing a non-Malthusian nature with the assistance of Darwinian principles. Kropotkin felt morally outraged by Huxley's Malthusian essay, "The Struggle for Existence in Human Society" (1888), in which Huxley had argued that the natural state for all organisms was a Hobbesian war of each against all. "From the point of view of the moralist," Huxley had argued, "the animal world is on about the same level as a gladiator's show . . . the strongest, the swiftest, and the cunningest live to fight another day . . . [and] no quarter is given" (1893–94, 199–200).

Kropotkin's response—eight articles written between 1890 and 1896 on collaboration as a universal adaptive strategy—was collected in 1902 under the title *Mutual Aid: A Factor of Evolution*. Recounting how during his own extensive travels in Siberia he had witnessed not struggle but solidarity among organisms, Kropotkin drew on Darwin's thesis of the adaptive function of intraspecies cooperation. Kropotkin contested Huxley's—and Darwin's own—Malthusian assumption that in nature the combined destabilizing pressures of population growth and variability in inheritance tended to force species into a natural state of individualistic, aggressive self-aggrandizement. Who are the fittest? Kropotkin asked: those continually at war with each other, or those who support each other? (1914, 14). Darwin's references to individual struggle, Kropotkin argued, were mainly metaphorical, for Darwin had produced little empirical evidence of severe struggles among individuals and varieties of the same species.

Kropotkin's ingenious blend of biological and social discourse, developed in defense of a radical laissez-faire theory of human society, reanimated the romantic view of primitive human innocence and benevolence. Prehistoric society was a state of "primitive communism" and "tribal solidarity" later corrupted by the development of the modern nuclear family (ibid., 94–97). Thus, Kropotkin read community, not individualism, from the natural state. He exaggerated Darwinian struggle—and Hobbesian war—to mean continual, visible confrontation, rather than the more subtle but still fatal strategies by which some organisms dominated food supply and breeding privilege. Hobbes had not meant by *war* literal battle among individuals or groups. "For Warre," he had insisted, "consisteth not in Battell onely, or the act of fighting; but in a tract of time, wherein the Will to contend by Battell is sufficiently known. . . . So the nature of War, consisteth not in actuall fighting; but in the known disposition thereto, during all the time there is no assurance to the contrary" (Hobbes 1651, 1:185–86). War did not mean for Hobbes, just as struggle did not mean for Darwin, literal confrontation to the death, but rather the disposition to use force as a utilitarian strategy of securing an individual end.

Despite Kropotkin, Huxley's confrontationist themes were taken up with renewed force by his former protégé E. Ray Lankester, the Oxford morphologist, in the Romanes Lecture of 1905, titled "Nature's Insurgent Son." Lankester, president of the British Association for the Advancement of Science and an intimate of H. G. Wells, Huxley's former student, hammered on the theme, consistent with Marxist ideology, that "the knowledge and control of nature is Man's destiny and his greatest need" (1907, 60). Author of a volume on *Extinct Animals* (1905), Lankester referred to the "unobtrusive, yet tremendous slaughter of the unfit which is incessantly going on" (ibid., 13). Reiterating some of the themes of his earlier volume, *Degeneration*, he argued that humans had begun a process of revolt against nature from which they could no longer withdraw without immense and catastrophic consequences to civilization.

Huxley's rejection of strict ethical utilitarianism opened up questions that continued to be debated in the terms of *Evolution and Ethics*. Some, like George E. Moore, sought to weaken what they saw as the metaphorical authority Victorian biology had assumed over ethics. In his *Principia Ethica* (1903), for example, Moore explored the so-called naturalistic fallacy, by which Bentham and Spencer had identified the good with natural process and pleasure (Moore 1903, 48–50; see also Huxley, p. 74). The *facts* of evolution were assumed to be the rules of natural process and, by theodicy, of ethical behavior. Pointing to the fallacy that "more evolved" meant "higher," he observed, as Huxley had, that an alteration in the earth's conditions would mechanically change the significance of "more evolved" (Moore 1903, 47; Huxley, pp. 4–5, 80–81; see also Richards 1987, 323–25).

Yet the rapidly developing fields of behaviorism and psychology could draw on a body of developing knowledge that had far more subtle explanatory potential than Spencer's simple analogical naturalism, as both Ruse and Richards have attempted to demonstrate (Ruse 1986; Richards 1987). In "Evolutionary Ethics," his own Romanes Lecture of 1943, Julian Huxley patronizingly dismissed his grandfather's misplaced Victorian moral standards as no longer self-evident. Sounding rather like Spencer, Julian Huxley sought to unify cosmic and ethical process by once more causally subordinating ethics to natural principles (Huxley and Huxley, 1947, 131; see Toulmin 1982, 53–71). Rejecting intuitive theories of the good, he declared ethics to be a loose term for conventions of rightness and wrongness. "Ethical realism" was the key, the degree to which ethical judgments conformed to external facts of experience. The greater the knowledge, the greater the degree of realism and good. Even Nazi indulgence of aggression and violence towards Jews was an example of ethical behavior, albeit a "grossly unrealistic one" (Huxley 1947, 123). Evolutionary advantages accrued to groups that could pool experience and coordinate actions that were based on a higher order of reality.

In the reception of Huxley's *Evolution and Ethics*, we find an intellectual and social divide. Older Victorian ethical naturalisms, including those of Spencer, Darwin, Clifford, and Stephen, tended to locate a direct operational analogy for ethical behavior in evolutionary utility. That is, the evolutionary process was conceived as the model of the ethical process. Individuals with these views sought explanatory parallels between human and animal behavior and saw Huxley's Romanes Lecture as dualistic and therefore inconsistent with his own naturalism. In a sense, they were right, for Huxley had once been a partisan of such naive analogical naturalism. On the other hand, Huxley did not reject evolutionary principles as the basis for the emergence of morality. Rather he concluded that evolutionary principles were not themselves to be taken as ethical principles.

CONCLUSION: THE HUMAN COLONY

Huxley's Romanes Lecture was pronounced "one of the most brilliant gems in the prose literature of the nineteenth century" by the *Oxford Magazine* ("Evolution," 1893b). Yet it was an essay about limitation and failure: the profound physical limits experienced by the human organism, and the failure of simple naturalism to light the way through the gathering social conflicts of the Victorian twilight. If Huxley self-critically pondered the metaphorical excesses of the century's biological speculators, his haunting image of human isolation remained the naturalistic product of Victorian scientific cosmology. But now the human colony, contending against entropy and instinct in an alien cosmic milieu, could seek little reassurance in the once-familiar equation "follow nature" (see Meyers 1985). Huxley had few answers, concluding his essay conservatively with a wistful backward glance at Tennyson's "Ulysses," and calling upon his fellows "To play the man 'strong in will / To strive, to seek, to find, and not to yield' " (p. 86). The Victorian ship was setting sail.

In the sequestered human culture Huxley imagined as a colony apart from the natural realm, we find one extreme of the secularization of human ends that had so tortured the nineteenth-century sensibility. The immediate ends of the individual and contemporary society had effectively displaced from Huxley's cultural agenda all theoretical futures—whether Christian, pantheist, evolutionary, or Marxist. "It is not clear," he had observed sarcastically in 1888, "what compensation the *Eohippus* gets for his sorrows in the fact that, some millions of years afterwards, one of his descendents wins the Derby" (1893–94, 9:199). Human purpose, if partly determined by instinct, had through consciousness and culture taken as its object the shaping of human conditions of existence according to an ethical ideal. But there were no teleological consolations, no evidences of a design, no unfolding necessities in nature, no meaningful hopes of a dramatically higher human destiny. Huxley adopted, rather, the practical demotic object of constructing by human artifice more humane, if necessarily temporary, terms of existence. This isolationist view of the human condition was a Victorian product that was to become familiar as a modern version of the cosmological picture, haunted by the prospect of Heat Death and isolated consciousness (Barrow and Tipler 1986).

Isolationism meant for Huxley the collapse of the explanatory biological analogy he had once championed with Spencer, Stephen, and other naturalists. Such Victorian analogical naturalism, as epitomized in Spencer's work, had endeavored to explain complex social processes by reference to specific physiological, embryological, and evolutionary processes. Even Darwin, somewhat in spite of himself, had thought of the increasing complexity of species as a natural progress toward what was innately *higher* rather than simply better suited to current conditions of existence (Ospovat 1981, 228). For Huxley, this effort to derive human terms of social existence from evolutionary process had come to seem a dangerous exercise. "It is very desirable," he now observed in his "Prolegomena," "to remember that evolution is not an explanation of the cosmic process, but merely a generalized statement of the method and results of that process" (p. 6). No longer did the evolutionary transformation of humankind loom as an end immanent in nature, with a theodical basis in the argument that continuous transmutation supported a pervasive progressive momentum from the lower to the higher. In Huxley's *Evolution and Ethics*, the Victorian dream of an "ideal man" was thus discarded.

In rejecting naive analogical naturalism, Huxley did not abandon the empirical principles of scientific naturalism or the assumption that culture and mind were the natural products of some unknown physical process. The divided individual could be seen in the demonstrated conflict of consciousness with instinct. Huxley's dualism was thus naturalized in the psychological division that he now viewed as the fundamental truth of human existence. As a characteristic of the brain, such internal human conflict was for all practical purposes permanent. Hence, environment and society became the only remaining domains of human adjustment, a view comprehensively developed by Huxley's former associate Lloyd Morgan in *Habit and Instinct*:

> Evolution has *been transferred from the organism to his environment*. There must increment somewhere, otherwise evolution is impossible. In social evolution on this view, the increment is by storage in the social environment to which each new generation adapts itself, with no increased native power of adaptation. In the written record, in social traditions, in the manifold inventions which make scientific and industrial progress possible, in the products of art, and the recorded examples of noble lives, we have an environment which is at the same time the product of mental evolution, and affords the condition of the development of each individual mind to-day. (1896, 340)

This attractive yet problematic theme that humans can effectively influence their collective destiny by altering environment, so prominent in the thought of Wallace, Darwin, Clifford, Huxley, Freud, Whitehead, Dewey, Julian Huxley, and many others, was reinforced by the Baconian view of science as control. It is a theme that has found recent expression in René Dubois's *The Wooing of Earth* (1980).

In Huxley's two essays, Victorian technological man became at once the personification of artifice and the gardener-force of the evolutionary world view. Although he could not effectively select, his powers were the very greatest to which humans could aspire, for he could alter conditions of existence—the strong force of Darwin's evolutionary dynamic. This unprecedented human competency, identified with the man of science, whom Huxley had championed among Victorians, gave considerable social consistency to Huxley's lifelong efforts on behalf of science education. It also brought upon Huxley the charge of petty liberal elitism, of seeking the marriage of science and capital on behalf of the captains of industry (Helfand 1977). Yet, the political intent of Huxley's "Evolution and Ethics" and "Prolegomena," as was clear in previous essays, including "Government: Anarchy or Regimentation" (1893–94, 1:383–430), was to seek some political middle ground from which to advance common social causes and exploit the growing "dominion over Nature" (ibid., 1:423). To Huxley, socialism overlooked the more fundamental problem of Malthus in favor of the secondary problem of distribution; on the other hand, laissez-faire individualism acquiesced to the status quo (ibid., 1:427–28).

The successes of the industrial revolution had made possible a new nature, "begotten by science upon fact." Such Baconian transformism, made more forceful by the industrial revolution, was blind to the equilibrium of environment (see Leiss 1972; Worster 1977). Huxley's colony distilled from Victorian culture the virtues of self-sufficiency and human enterprise amid the alien internal and external forces of the physical world. But in its imperial and single-minded drive to neutralize the forces of the antagonist, it demonstrated little sensitivity to the dependencies of humankind in the fabric of nature. Huxley had little sense of the ecological whole. If he was more realistic in his views of human limits than many of his contemporaries, his program was also ambiguous, for it implied the apotheosis of control in a war against the self and nature that none dare win.

EVOLUTION & ETHICS

BY

THOMAS H. HUXLEY

London
MACMILLAN AND CO.
1894

I

EVOLUTION AND ETHICS

PROLEGOMENA

[1894]

I

IT may be safely assumed that, two thousand years ago, before Cæsar set foot in southern Britain, the whole country-side visible from the windows of the room in which I write, was in what is called " the state of nature." Except, it may be, by raising a few sepulchral mounds, such as those which still, here and there, break the flowing contours of the downs, man's hands had made no mark upon it ; and the thin veil of vegetation which overspread the broad-backed heights and the shelving sides of the coombs was unaffected by his industry. The native grasses and weeds, the scattered patches of gorse, contended with one another for the possession of the scanty surface soil ; they fought against the droughts of summer,

the frosts of winter, and the furious gales which
swept, with unbroken force, now from the Atlantic,
and now from the North Sea, at all times of the
year ; they filled up, as they best might, the gaps
made in their ranks by all sorts of underground
and overground animal ravagers. One year with
another, an average population, the floating balance
of the unceasing struggle for existence among the
indigenous plants, maintained itself. It is as
little to be doubted, that an essentially similar
state of nature prevailed, in this region, for many
thousand years before the coming of Cæsar ; and
there is no assignable reason for denying that it
might continue to exist through an equally prolonged futurity, except for the intervention of man.

Reckoned by our customary standards of
duration, the native vegetation, like the " everlasting hills " which it clothes, seems a type of
permanence. The little Amarella Gentians, which
abound in some places to-day, are the descendants
of those that were trodden underfoot by the prehistoric savages who have left their flint tools about,
here and there ; and they followed ancestors
which, in the climate of the glacial epoch, probably
flourished better than they do now.   Compared
with the long past of this humble plant, all the
history of civilized men is but an episode.

Yet nothing is more certain than that, measured
by the liberal scale of time-keeping of the universe,
this present state of nature, however it may seem

to have gone and to go on for ever, is but a fleeting phase of her infinite variety; merely the last of the series of changes which the earth's surface has undergone in the course of the millions of years of its existence. Turn back a square foot of the thin turf, and the solid foundation of the land, exposed in cliffs of chalk five hundred feet high on the adjacent shore, yields full assurance of a time when the sea covered the site of the " everlasting hills "; and when the vegetation of what land lay nearest, was as different from the present Flora of the Sussex downs, as that of Central Africa now is.[1] No less certain is it that, between the time during which the chalk was formed and that at which the original turf came into existence, thousands of centuries elapsed, in the course of which, the state of nature of the ages during which the chalk was deposited, passed into that which now is, by changes so slow that, in the coming and going of the generations of men, had such witnessed them, the contemporary conditions would have seemed to be unchanging and unchangeable.

But it is also certain that, before the deposition of the chalk, a vastly longer period had elapsed, throughout which it is easy to follow the traces of the same process of ceaseless modification and of the internecine struggle for existence of living things; and that even when we can get no further

[1] See "On a piece of Chalk" in the preceding volume of these Essays (vol. viii. p. 1).

back, it is not because there is any reason to think
we have reached the beginning, but because the
trail of the most ancient life remains hidden, or
has become obliterated.

Thus that state of nature of the world of plants,
which we began by considering, is far from possess-
ing the attribute of permanence. Rather its very
essence is impermanence. It may have lasted
twenty or thirty thousand years, it may last for
twenty or thirty thousand years more, without
obvious change; but, as surely as it has followed
upon a very different state, so it will be followed
by an equally different condition. That which
endures is not one or another association of living
forms, but the process of which the cosmos is the
product, and of which these are among the transitory expressions. And in the living world, one of
the most characteristic features of this cosmic process is the struggle for existence, the competition
of each with all, the result of which is the selection,
that is to say, the survival of those forms which,
on the whole, are best adapted to the conditions
which at any period obtain; and which are, therefore, in that respect, and only in that respect, the
fittest.[1] The acme reached by the cosmic process

[1] That every theory of evolution must be consistent not
merely with progressive development, but with indefinite
persistence in the same condition and with retrogressive modification, is a point which I have insisted upon repeatedly from
the year 1862 till now. See *Collected Essays*, vol. ii. pp. 461–89 ;
vol. iii. p. 33 ; vol. viii. p. 304. In the address on "Geological

in the vegetation of the downs is seen in the turf, with its weeds and gorse. Under the conditions, they have come out of the struggle victorious; and, by surviving, have proved that they are the fittest to survive.

That the state of nature, at any time, is a temporary phase of a process of incessant change, which has been going on for innumerable ages, appears to me to be a proposition as well established as any in modern history. Paleontology assures us, in addition, that the ancient philosophers who, with less reason, held the same doctrine, erred in supposing that the phases formed a cycle, exactly repeating the past, exactly foreshadowing the future, in their rotations. On the contrary, it furnishes us with conclusive reasons for thinking that, if every link in the ancestry of these humble indigenous plants had been preserved and were accessible to us, the whole would present a converging series of forms of gradually diminishing complexity, until, at some period in the history of the earth, far more remote than any of which organic remains have yet been discovered, they would merge in those low groups among which the boundaries between animal and vegetable life become effaced.[1]

Contemporaneity and Persistent Types " (1862), the paleontological proofs of this proposition were, I believe, first set forth.

[1] "On the Border Territory between the Animal and the Vegetable Kingdoms," Essays, vol. viii. p. 162.

The word " evolution," now generally applied to the cosmic process, has had a singular history, and is used in various senses.[1] Taken in its popular signification it means progressive development, that is, gradual change from a condition of relative uniformity to one of relative complexity; but its connotation has been widened to include the phenomena of retrogressive metamorphosis, that is, of progress from a condition of relative complexity to one of relative uniformity.

As a natural process, of the same character as the development of a tree from its seed, or of a fowl from its egg, evolution excludes creation and all other kinds of supernatural intervention. As the expression of a fixed order, every stage of which is the effect of causes operating according to definite rules, the conception of evolution no less excludes that of chance. It is very desirable to remember that evolution is not an explanation of the cosmic process, but merely a generalized statement of the method and results of that process. And, further, that, if there is proof that the cosmic process was set going by any agent, then that agent will be the creator of it and of all its products, although supernatural intervention may remain strictly excluded from its further course.

So far as that limited revelation of the nature of things, which we call scientific knowledge, has

[1] See " Evolution in Biology," Essays, vol. ii. p. 187.

yet gone, it tends, with constantly increasing emphasis, to the belief that, not merely the world of plants, but that of animals; not merely living things, but the whole fabric of the earth ; not merely our planet, but the whole solar system ; not merely our star and its satellites, but the millions of similar bodies which bear witness to the order which pervades boundless space, and has endured through boundless time ; are all working out their predestined courses of evolution.

With none of these have I anything to do, at present, except with that exhibited by the forms of life which tenant the earth. All plants and animals exhibit the tendency to vary, the causes of which have yet to be ascertained; it is the tendency of the conditions of life, at any given time, while favouring the existence of the variations best adapted to them, to oppose that of the rest and thus to exercise selection ; and all living things tend to multiply without limit, while the means of support are limited; the obvious cause of which is the production of offspring more numerous than their progenitors, but with equal expectation of life in the actuarial sense. Without the first tendency there could be no evolution. Without the second, there would be no good reason why one variation should disappear and another take its place; that is to say, there would be no selection. Without the third, the struggle for existence, the agent of the selective process in the state of nature, would vanish.[1]

Granting the existence of these tendencies, all the known facts of the history of plants and of animals may be brought into rational correlation. And this is more than can be said for any other hypothesis that I know of. Such hypotheses, for example, as that of the existence of a primitive, orderless chaos; of a passive and sluggish eternal matter moulded, with but partial success, by archetypal ideas; of a brand-new world-stuff suddenly created and swiftly shaped by a supernatural power; receive no encouragement, but the contrary, from our present knowledge. That our earth may once have formed part of a nebulous cosmic magma is certainly possible, indeed seems highly probable; but there is no reason to doubt that order reigned there, as completely as amidst what we regard as the most finished works of nature or of man.[2] The faith which is born of knowledge, finds its object in an eternal order, bringing forth ceaseless change, through endless time, in endless space; the manifestations of the cosmic energy alternating between phases of potentiality and phases of explication. It may be that, as Kant suggests,[3] every cosmic

[1] *Collected Essays*, vol. ii. *passim*.
[2] *Ibid.*, vol. iv. p. 138 ; vol. v. pp. 71-73.
[3] *Ibid.*, vol. viii. p. 321.

magma predestined to evolve into a new world, has been the no less predestined end of a vanished predecessor.

II

Three or four years have elapsed since the state of nature, to which I have referred, was brought to an end, so far as a small patch of the soil is concerned, by the intervention of man. The patch was cut off from the rest by a wall; within the area thus protected, the native vegetation was, as far as possible, extirpated; while a colony of strange plants was imported and set down in its place. In short, it was made into a garden. At the present time, this artificially treated area presents an aspect extraordinarily different from that of so much of the land as remains in the state of nature, outside the wall. Trees, shrubs, and herbs, many of them appertaining to the state of nature of remote parts of the globe, abound and flourish. Moreover, considerable quantities of vegetables, fruits, and flowers are produced, of kinds which neither now exist, nor have ever existed, except under conditions such as obtain in the garden; and which, therefore, are as much works of the art of man as the frames and glass-houses in which some of them are raised. That the "state of Art," thus created in the state of nature by man, is sustained by and dependent on him, would at once become apparent, if the watchful supervision of the gardener were withdrawn, and the antagonistic influences of the general cosmic process were no longer sedulously warded off, or counteracted. The walls and gates would decay; quadrupedal and bipedal intruders would devour and tread down the useful and beautiful plants; birds, insects, blight, and mildew would work their will; the seeds of the native plants, carried by winds or other agencies, would immigrate, and in virtue of their long-earned special adaptation to the local conditions, these despised native weeds would soon choke their choice exotic rivals. A century or two hence, little beyond the foundations of the wall and of the houses and frames would be left, in evidence of the victory of the cosmic powers at work in the state of nature, over the temporary obstacles to their supremacy, set up by the art of the horticulturist.

It will be admitted that the garden is as much a work of art,[1] or artifice, as anything that can be mentioned. The energy localised in certain human bodies, directed by similarly localised intellects, has produced a collocation of other material bodies which could not be brought about in the state of nature. The same proposition is true of all the

---

[1] The sense of the term "Art" is becoming narrowed; "work of Art" to most people means a picture, a statue, or a piece of *bijouterie*; by way of compensation "artist" has included in its wide embrace cooks and ballet girls, no less than painters and sculptors.

works of man's hands, from a flint implement to a cathedral or a chronometer; and it is because it is true, that we call these things artificial, term them works of art, or artifice, by way of distinguishing them from the products of the cosmic process, working outside man, which we call natural, or works of nature. The distinction thus drawn between the works of nature and those of man, is universally recognised; and it is, as I conceive, both useful and justifiable.

III

No doubt, it may be properly urged that the operation of human energy and intelligence, which has brought into existence and maintains the garden, by what I have called "the horticultural process," is, strictly speaking, part and parcel of the cosmic process. And no one could more readily agree to that proposition than I. In fact, I do not know that any one has taken more pains than I have, during the last thirty years, to insist upon the doctrine, so much reviled in the early part of that period, that man, physical, intellectual, and moral, is as much a part of nature, as purely a product of the cosmic process, as the humblest weed.[1]

But if, following up this admission, it is urged

---

[1] See "Man's Place in Nature," *Collected Essays*, vol. vii., and "On the Struggle for Existence in Human Society" (1888), below.

that, such being the case, the cosmic process cannot be in antagonism with that horticultural process which is part of itself—I can only reply, that
if the conclusion that the two are antagonistic
is logically absurd, I am sorry for logic, because,
as we have seen, the fact is so. The garden is in
the same position as every other work of man's
art; it is a result of the cosmic process working
through and by human energy and intelligence ;
and, as is the case with every other artificial
thing set up in the state of nature, the influences of the latter are constantly tending to break
it down and destroy it. No doubt, the Forth bridge
and an ironclad in the offing, are, in ultimate re-
sort, products of the cosmic process ; as much so as
the river which flows under the one, or the sea-
water on which the other floats. Nevertheless,
every breeze strains the bridge a little, every tide
does something to weaken its foundations ; every
change of temperature alters the adjustment of
its parts, produces friction and consequent wear
and tear. From time to time, the bridge must be
repaired, just as the ironclad must go into dock ;
simply because nature is always tending to re-
claim that which her child, man, has borrowed
from her and has arranged in combinations which
are not those favoured by the general cosmic
process.

Thus, it is not only true that the cosmic
energy, working through man upon a portion of the

plant world, opposes the same energy as it works through the state of nature, but a similar antagonism is everywhere manifest between the artificial and the natural. Even in the state of nature itself, what is the struggle for existence but the antagonism of the results of the cosmic process in the region of life, one to another ? [1]

IV

Not only is the state of nature hostile to the state of art of the garden; but the principle of the horticultural process, by which the latter is created and maintained, is antithetic to that of the cosmic process. The characteristic feature of the latter is the intense and unceasing competition of the struggle for existence. The characteristic of the former is the elimination of that struggle, by the removal of the conditions which give rise to it. The tendency of the cosmic process is to bring about the adjustment of the forms of plant life to the current conditions; the tendency of the horticultural process is the adjustment of the conditions to the needs of the forms of plant life which the gardener desires to raise.

The cosmic process uses unrestricted multiplica-

---

[1] Or to put the case still more simply. When a man lays hold of the two ends of a piece of string and pulls them, with intent to break it, the right arm is certainly exerted in antagonism to the left arm ; yet both arms derive their energy from the same original source.

tion as the means whereby hundreds compete for the place and nourishment adequate for one; it employs frost and drought to cut off the weak and unfortunate; to survive, there is need not only of strength, but of flexibility and of good fortune.

The gardener, on the other hand, restricts multiplication; provides that each plant shall have sufficient space and nourishment; protects from frost and drought; and, in every other way, attempts to modify the conditions, in such a manner as to bring about the survival of those forms which most nearly approach the standard of the useful, or the beautiful, which he has in his mind.

If the fruits and the tubers, the foliage and the flowers thus obtained, reach, or sufficiently approach, that ideal, there is no reason why the *status quo* attained should not be indefinitely prolonged. So long as the state of nature remains approximately the same, so long will the energy and intelligence which created the garden suffice to maintain it. However, the limits within which this mastery of man over nature can be maintained are narrow. If the conditions of the cretaceous epoch returned, I fear the most skilful of gardeners would have to give up the cultivation of apples and gooseberries; while, if those of the glacial period once again obtained, open asparagus beds would be superfluous, and the training of fruit trees against the most favourable of south walls, a waste of time and trouble.

But it is extremely important to note that, the state of nature remaining the same, if the produce does not satisfy the gardener, it may be made to approach his ideal more closely. Although the struggle for existence may be at end, the possibility of progress remains. In discussions on these topics, it is often strangely forgotten that the essential conditions of the modification, or evolution, of living things are variation and hereditary transmission. Selection is the means by which certain variations are favoured and their progeny preserved. But the struggle for existence is only one of the means by which selection may be effected. The endless varieties of cultivated flowers, fruits, roots, tubers, and bulbs are not products of selection by means of the struggle for existence, but of direct selection, in view of an ideal of utility or beauty. Amidst a multitude of plants, occupying the same station and subjected to the same conditions, in the garden, varieties arise. The varieties tending in a given direction are preserved, and the rest are destroyed. And the same process takes place among the varieties until, for example, the wild kale becomes a cabbage, or the wild *Viola tricolor* a prize pansy.

V

The process of colonization presents analogies to the formation of a garden which are highly instructive. Suppose a shipload of English colonists sent to form a settlement, in such a country as Tasmania was in the middle of the last century. On landing, they find themselves in the midst of a state of nature, widely different from that left behind them in everything but the most general physical conditions. The common plants, the common birds and quadrupeds, are as totally distinct as the men from anything to be seen on the side of the globe from which they come. The colonists proceed to put an end to this state of things over as large an area as they desire to occupy. They clear away the native vegetation, extirpate or drive out the animal population, so far as may be necessary, and take measures to defend themselves from the re-immigration of either. In their place, they introduce English grain and fruit trees; English dogs, sheep, cattle, horses; and English men; in fact, they set up a new Flora and Fauna and a new variety of mankind, within the old state of nature. Their farms and pastures represent a garden on a great scale, and themselves the gardeners who have to keep it up, in watchful antagonism to the old *régime*. Considered as a whole, the colony is a composite unit introduced into the old state of nature; and, thenceforward, a competitor in the struggle for existence, to conquer or be vanquished.

Under the conditions supposed, there is no doubt of the result, if the work of the colonists be carried out energetically and with intelligent combination of all their forces. On the other hand, if they are slothful, stupid, and careless ; or if they waste their energies in contests with one another, the chances are that the old state of nature will have the best of it. The native savage will destroy the immigrant civilized man ; of the English animals and plants some will be extirpated by their indigenous rivals, others will pass into the feral state and themselves become components of the state of nature. In a few decades, all other traces of the settlement will have vanished.

VI

Let us now imagine that some administrative authority, as far superior in power and intelligence to men, as men are to their cattle, is set over the colony, charged to deal with its human elements in such a manner as to assure the victory of the settlement over the antagonistic influences of the state of nature in which it is set down. He would proceed in the same fashion as that in which the gardener dealt with his garden. In the first place, he would, as far as possible, put a stop to the influence of external competition by thoroughly extirpating and excluding the native rivals, whether men, beasts, or plants. And our administrator would select his human agents, with a view to his ideal of a successful colony, just as the gardener selects his plants with a view to his ideal of useful or beautiful products.

In the second place, in order that no struggle for the means of existence between these human agents should weaken the efficiency of the corporate whole in the battle with the state of nature, he would make arrangements by which each would be provided with those means; and would be relieved from the fear of being deprived of them by his stronger or more cunning fellows. Laws, sanctioned by the combined force of the colony, would restrain the self-assertion of each man within the limits required for the maintenance of peace. In other words, the cosmic struggle for existence, as between man and man, would be rigorously suppressed; and selection, by its means, would be as completely excluded as it is from the garden.

At the same time, the obstacles to the full development of the capacities of the colonists by other conditions of the state of nature than those already mentioned, would be removed by the creation of artificial conditions of existence of a more favourable character. Protection against extremes of heat and cold would be afforded by houses and clothing; drainage and irrigation works would antagonise the effects of excessive rain and excessive drought; roads, bridges, canals, carriages, and ships would overcome the natural obstacles to locomotion and transport; mechanical engines would supplement the natural strength of men and of their draught animals; hygienic precautions would check, or remove, the natural causes of disease. With every step of this progress in civilization, the colonists would become more and more independent of the state of nature; more and more, their lives would be conditioned by a state of art. In order to attain his ends, the administrator would have to avail himself of the courage, industry, and co-operative intelligence of the settlers; and it is plain that the interest of the community would be best served by increasing the proportion of persons who possess such qualities, and diminishing that of persons devoid of them. In other words, by selection directed towards an ideal.

Thus the administrator might look to the establishment of an earthly paradise, a true garden of Eden, in which all things should work together towards the well-being of the gardeners: within which the cosmic process, the coarse struggle for existence of the state of nature, should be abolished; in which that state should be replaced by a state of art;

where every plant and every lower animal should be adapted to human wants, and would perish if human supervision and protection were withdrawn; where men themselves should have been selected, with a view to their efficiency as organs for the performance of the functions of a perfected society. And this ideal polity would have been brought about, not by gradually adjusting the men to the conditions around them, but by creating artificial conditions for them; not by allowing the free play of the struggle for existence, but by excluding that struggle; and by substituting selection directed towards the administrator's ideal for the selection it exercises.

VII

But the Eden would have its serpent, and a very subtle beast too. Man shares with the rest of the living world the mighty instinct of reproduction and its consequence, the tendency to multiply with great rapidity. The better the measures of the administrator achieved their object, the more completely the destructive agencies of the state of nature were defeated, the less would that multiplication be checked.

On the other hand, within the colony, the enforcement of peace, which deprives every man of the power to take away the means of existence from another, simply because he is the stronger,

would have put an end to the struggle for existence between the colonists, and the competition for
the commodities of existence, which would alone
remain, is no check upon population.

Thus, as soon as the colonists began to multiply,
the administrator would have to face the tendency
to the reintroduction of the cosmic struggle into
his artificial fabric, in consequence of the competition, not merely for the commodities, but for the
means of existence. When the colony reached
the limit of possible expansion, the surplus population must be disposed of somehow; or the fierce
struggle for existence must recommence and
destroy that peace, which is the fundamental condition of the maintenance of the state of art
against the state of nature.

Supposing the administrator to be guided by
purely scientific considerations, he would, like the
gardener, meet this most serious difficulty by
systematic extirpation, or exclusion, of the superfluous. The hopelessly diseased, the infirm aged,
the weak or deformed in body or in mind, the
excess of infants born, would be put away, as the
gardener pulls up defective and superfluous plants,
or the breeder destroys undesirable cattle. Only
the strong and the healthy, carefully matched, with
a view to the progeny best adapted to the purposes of the administrator, would be permitted to
perpetuate their kind.

VIII

Of the more thoroughgoing of the multitudinous attempts to apply the principles of cosmic evolution, or what are supposed to be such, to social and political problems, which have appeared of late years, a considerable proportion appear to me to be based upon the notion that human society is competent to furnish, from its own resources, an administrator of the kind I have imagined. The pigeons, in short, are to be their own Sir John Sebright.[1] A despotic government, whether individual or collective, is to be endowed with the preternatural intelligence, and with what, I am afraid, many will consider the preternatural ruthlessness, required for the purpose of carrying out the principle of improvement by selection, with the somewhat drastic thoroughness upon which the success of the method depends. Experience certainly does not justify us in limiting the ruthlessness of individual "saviours of society"; and, on the well-known grounds of the aphorism which denies both body and soul to corporations, it seems probable (indeed the belief is not without support in history) that a collective despotism, a mob got to believe in its own divine right by demagogic missionaries, would be capable of more thorough

[1] Not that the conception of such a society is necessarily based upon the idea of evolution. The Platonic state testifies to the contrary.

work in this direction than any single tyrant, puffed
up with the same illusion, has ever achieved.
But intelligence is another affair. The fact that
" saviours of society " take to that trade is evidence
enough that they have none to spare. And
such as they possess is generally sold to the
capitalists of physical force on whose resources
they depend. However, I doubt whether even
the keenest judge of character, if he had before
him a hundred boys and girls under fourteen, could
pick out, with the least chance of success, those
who should be kept, as certain to be serviceable
members of the polity, and those who should be
chloroformed, as equally sure to be stupid, idle, or
vicious. The " points " of a good or of a bad citizen are really far harder to discern than those of
a puppy or a short-horn calf ; many do not show
themselves before the practical difficulties of life
stimulate manhood to full exertion. And by that
time the mischief is done. The evil stock, if it be
one, has had time to multiply, and selection is
nullified.

IX

I have other reasons for fearing that this
logical ideal of evolutionary regimentation—this
pigeon-fanciers' polity—is unattainable. In the
absence of any such a severely scientific administrator as we have been dreaming of, human society

is kept together by bonds of such a singular character, that the attempt to perfect society after his fashion would run serious risk of loosening them.

Social organization is not peculiar to men. Other societies, such as those constituted by bees and ants, have also arisen out of the advantage of co-operation in the struggle for existence; and their resemblances to, and their differences from, human society are alike instructive. The society formed by the hive bee fulfils the ideal of the communistic aphorism "to each according to his needs, from each according to his capacity." Within it, the struggle for existence is strictly limited. Queen, drones, and workers have each their allotted sufficiency of food; each performs the function assigned to it in the economy of the hive, and all contribute to the success of the whole co-operative society in its competition with rival collectors of nectar and pollen and with other enemies, in the state of nature without. In the same sense as the garden, or the colony, is a work of human art, the bee polity is a work of apiarian art, brought about by the cosmic process, working through the organization of the hymenopterous type.

Now this society is the direct product of an organic necessity, impelling every member of it to a course of action which tends to the good of the whole. Each bee has its duty and none has any rights. Whether bees are susceptible of feeling and capable of thought is a question which cannot be dogmatically answered. As a pious opinion, I am disposed to deny them more than the merest rudiments of consciousness.[1] But it is curious to reflect that a thoughtful drone (workers and queens would have no leisure for speculation) with a turn for ethical philosophy, must needs profess himself an intuitive moralist of the purest water. He would point out, with perfect justice, that the devotion of the workers to a life of ceaseless toil for a mere subsistence wage, cannot be accounted for either by enlightened selfishness, or by any other sort of utilitarian motives; since these bees begin to work, without experience or reflection, as they emerge from the cell in which they are hatched. Plainly, an eternal and immutable principle, innate in each bee, can alone account for the phenomena. On the other hand, the biologist, who traces out all the extant stages of gradation between solitary and hive bees, as clearly sees in the latter, simply the perfection of an automatic mechanism, hammered out by the blows of the struggle for existence upon the progeny of the former, during long ages of constant variation.

[1] *Collected Essays*, vol. i., "Animal Automatism"; vol. v., "Prologue," pp. 45 *et seq.*

X

I see no reason to doubt that, at its origin, human society was as much a product of organic necessity as that of the bees.[1] The human family, to begin with, rested upon exactly the same conditions as those which gave rise to similar associations among animals lower in the scale. Further, it is easy to see that every increase in the duration of the family ties, with the resulting co-operation of a larger and larger number of descendants for protection and defence, would give the families in which such modification took place a distinct advantage over the others. And, as in the hive, the progressive limitation of the struggle for existence between the members of the family would involve increasing efficiency as regards outside competition.

But there is this vast and fundamental difference between bee society and human society. In the former, the members of the society are each organically predestined to the performance of one particular class of functions only. If they were endowed with desires, each could desire to perform none but those offices for which its organization specially fits it; and which, in view of the good of the whole, it is proper it should do. So long as a new queen does not make her appearance, rivalries and competition are absent from the bee polity.

[1] *Collected Essays*, vol. v., Prologue, pp. 50-54.

Among mankind, on the contrary, there is no such predestination to a sharply defined place in the social organism. However much men may differ in the quality of their intellects, the intensity of their passions, and the delicacy of their sensations, it cannot be said that one is fitted by his organization to be an agricultural labourer and nothing else, and another to be a landowner and nothing else. Moreover, with all their enormous differences in natural endowment, men agree in one thing, and that is their innate desire to enjoy the pleasures and to escape the pains of life; and, in short, to do nothing but that which it pleases them to do, without the least reference to the welfare of the society into which they are born. That is their inheritance (the reality at the bottom of the doctrine of original sin) from the long series of ancestors, human and semi-human and brutal, in whom the strength of this innate tendency to self-assertion was the condition of victory in the struggle for existence. That is the reason of the *aviditas vitæ*[1]—the insatiable hunger for enjoyment—of all mankind, which is one of the essential conditions of success in the war with the state of nature outside; and yet the sure agent of the destruction of society if allowed free play within.

The check upon this free play of self-assertion, or natural liberty, which is the necessary condition for the origin of human society, is the product

[1] See below. Romanes' Lecture, note 7.

of organic necessities of a different kind from
those upon which the constitution of the hive
depends. One of these is the mutual affection
of parent and offspring, intensified by the long
infancy of the human species. But the most
important is the tendency, so strongly
developed in man, to reproduce in himself actions and feelings similar to, or correlated with,
those of other men. Man is the most consummate of all mimics in the animal world;
none but himself can draw or model; none comes
near him in the scope, variety, and exactness of
vocal imitation; none is such a master of gesture;
while he seems to be impelled thus to imitate
for the pure pleasure of it. And there is
no such another emotional chameleon. By a
purely reflex operation of the mind, we take
the hue of passion of those who are about us,
or, it may be, the complementary colour. It is
not by any conscious " putting one's self in the
place " of a joyful or a suffering person that the
state of mind we call sympathy usually arises;[1]
indeed, it is often contrary to one's sense of

[1] Adam Smith makes the pithy observation that the man
who sympathises with a woman in childbed, cannot be said
to put himself in her place. ("The Theory of the Moral Sentiments," Part vii. sec. iii. chap. i.) Perhaps there is more
humour than force in the example; and, in spite of this
and other observations of the same tenor, I think that the
one defect of the remarkable work in which it occurs is that
it lays too much stress on conscious substitution, too little
on purely reflex sympathy.

right, and in spite of one's will, that "fellow-feeling makes us wondrous kind," or the reverse. However complete may be the indifference to public opinion, in a cool, intellectual view, of the traditional sage, it has not yet been my fortune to meet with any actual sage who took its hostile manifestations with entire equanimity. Indeed, I doubt if the philosopher lives, or ever has lived, who could know himself to be heartily despised by a street boy without some irritation. And, though one cannot justify Haman for wishing to hang Mordecai on such a very high gibbet, yet, really, the consciousness of the Vizier of Ahasuerus, as he went in and out of the gate, that this obscure Jew had no respect for him, must have been very annoying.[1]

It is needful only to look around us, to see that the greatest restrainer of the anti-social tendencies of men is fear, not of the law, but of the opinion of their fellows. The conventions of honour bind men who break legal, moral, and religious bonds; and, while people endure the extremity of physical pain rather than part with life, shame drives the weakest to suicide.

Every forward step of social progress brings men

[1] Esther v. 9-13. ". . . but when Haman saw Mordecai in the king's gate, that he stood not up, nor moved for him, he was full of indignation against Mordecai. . . . And Haman told them of the glory of his riches. . . . and all the things wherein the king had promoted him. . . . Yet all this availeth me nothing, so long as I see Mordecai the Jew sitting at the king's gate." What a shrewd exposure of human weakness it is !

into closer relations with their fellows, and increases the importance of the pleasures and pains
derived from sympathy. We judge the acts of
others by our own sympathies, and we judge our
own acts by the sympathies of others, every day
and all day long, from childhood upwards, until
associations, as indissoluble as those of language,
are formed between certain acts and the feelings
of approbation or disapprobation. It becomes
impossible to imagine some acts without disapprobation, or others without approbation of
the actor, whether he be one's self, or any one else.
We come to think in the acquired dialect of morals.
An artificial personality, the "man within," as
Adam Smith [1] calls conscience, is built up beside
the natural personality.   He is the watchman of
society, charged to restrain the anti-social tendencies of the natural man within the limits
required by social welfare.

XI

I have termed this evolution of the feelings
out of which the primitive bonds of human
society are so largely forged, into the organized
and personified sympathy we call conscience,
the ethical process.[2] So far as it tends to

[1] "Theory of the Moral Sentiments," Part iii. chap. 3.
On the influence and authority of conscience.

[2] Worked out, in its essential features, chiefly by Hartley and
Adam Smith, long before the modern doctrine of evolution was
thought of. See *Note* below, p. 45.

make any human society more efficient in the struggle for existence with the state of nature, or with other societies, it works in harmonious contrast with the cosmic process. But it is none the less true that, since law and morals are restraints upon the struggle for existence between men in society, the ethical process is in opposition to the principle of the cosmic process, and tends to the suppression of the qualities best fitted for success in that struggle.[1]

It is further to be observed that, just as the self-assertion, necessary to the maintenance of society against the state of nature, will destroy that society if it is allowed free operation within; so the self-restraint, the essence of the ethical process, which is no less an essential condition of the existence of every polity, may, by excess, become ruinous to it.

Moralists of all ages and of all faiths, attending only to the relations of men towards one another in an ideal society, have agreed upon the "golden rule," "Do as you would be done by." In other words, let sympathy be your guide; put yourself in the place of the man towards whom your action is directed; and do to him what you would like to have done to yourself under the circumstances. However much one may admire the generosity of such a rule of con-

---

[1] See the essay "On the Struggle for Existence in Human Society" below; and *Collected Essays*, vol. i. p. 276, for Kant's recognition of these facts.

duct; however confident one may be that average
men may be thoroughly depended upon not to
carry it out to its full logical consequences; it
is nevertheless desirable to recognise the fact
that these consequences are incompatible with
the existence of a civil state, under any circumstances of this world which have obtained, or,
so far as one can see, are, likely to come to
pass.

For I imagine there can be no doubt that the
great desire of every wrongdoer is to escape from
the painful consequences of his actions. If I put
myself in the place of the man who has robbed me,
I find that I am possessed by an exceeding desire
not to be fined or imprisoned; if in that of the
man who has smitten me on one cheek, I contemplate with satisfaction the absence of any worse
result than the turning of the other cheek for like
treatment. Strictly observed, the " golden rule "
involves the negation of law by the refusal to put it
in motion against law-breakers; and, as regards
the external relations of a polity, it is the refusal
to continue the struggle for existence. It can be
obeyed, even partially, only under the protection of
a society which repudiates it. Without such
shelter, the followers of the " golden rule " may in-
dulge in hopes of heaven, but they must reckon
with the certainty that other people will be
masters of the earth.

What would become of the garden if the gardener treated all the weeds and slugs and birds and trespassers as he would like to be treated, if he were in their place ?

XII

Under the preceding heads, I have endeavoured to represent in broad, but I hope faithful, outlines the essential features of the state of nature and of that cosmic process of which it is the outcome, so far as was needful for my argument ; I have contrasted with the state of nature the state of art, produced by human intelligence and energy, as it is exemplified by a garden ; and I have shown that the state of art, here and elsewhere, can be maintained only by the constant counteraction of the hostile influences of the state of nature. Further, I have pointed out that the " horticultural process " which thus sets itself against the " cosmic process " is opposed to the latter in principle, in so far as it tends to arrest the struggle for existence, by restraining the multiplication which is one of the chief causes of that struggle, and by creating artificial conditions of life, better adapted to the cultivated plants than are the conditions of the state of nature. And I have dwelt upon the fact that, though the progressive modification, which is the consequence of the struggle for existence in the state of nature, is at an end, such modification may still be effected by that selection, in view of an ideal of usefulness, or of pleasantness, to man, of which the state of nature knows nothing.

I have proceeded to show that a colony, set down in a country in the state of nature, presents close analogies with a garden; and I have indicated the course of action which an administrator, able and willing to carry out horticultural principles, would adopt, in order to secure the success of such a newly formed polity, supposing it to be capable of indefinite expansion. In the contrary case, I have shown that difficulties must arise ; that the unlimited increase of the population over a limited area must, sooner or later, reintroduce into the colony that struggle for the means of existence between the colonists, which it was the primary object of the administrator to exclude, insomuch as it is fatal to the mutual peace which is the prime condition of the union of men in society.

I have briefly described the nature of the only radical cure, known to me, for the disease which would thus threaten the existence of the colony ; and, however regretfully, I have been obliged to admit that this rigorously scientific method of applying the principles of evolution to human society hardly comes within the region of practical politics ; not for want of will on the part of a great many people; but because, for one reason, there is no hope that mere human beings will ever possess enough intelligence to select the fittest. And I have adduced other grounds for arriving at the same conclusion.

I have pointed out that human society took its rise in the organic necessities expressed by imitation and by the sympathetic emotions; and that, in the struggle for existence with the state of nature and with other societies, as part of it, those in which men were thus led to close co-operation had a great advantage.[1] But, since each man retained more or less of the faculties common to all the rest, and especially a full share of the desire for unlimited self-gratification, the struggle for existence within society could only be gradually eliminated. So long as any of it remained, society continued to be an imperfect instrument of the struggle for existence and, consequently, was improvable by the selective influence of that struggle. Other things being alike, the tribe of savages in which order was best maintained; in which there was most security within the tribe and the most loyal mutual support outside it, would be the survivors.

I have termed this gradual strengthening of the social bond, which, though it arrests the struggle for existence inside society, up to a certain point improves the chances of society, as a corporate whole, in the cosmic struggle—the ethical process. I have endeavoured to show that, when the ethical process has advanced so

[1] *Collected Essays,* vol. v., Prologue, p. 52.

far as to secure every member of the society in the possession of the means of existence, the struggle for existence, as between man and man, within that society is, *ipso facto*, at an end. And, as it is undeniable that the most highly civilized societies have substantially reached this position, it follows that, so far as they are concerned, the struggle for existence can play no important part within them.[1] In other words, the kind of evolution which is brought about in the state of nature cannot take place.

I have further shown cause for the belief that direct selection, after the fashion of the horticulturist and the breeder, neither has played, nor can play, any important part in the evolution of society ; apart from other reasons, because I do not see how such selection could be practised without a serious weakening, it may be the destruction, of the bonds which hold society together. It strikes me that men who are accustomed to contemplate the active or passive extirpation of the weak, the unfortunate, and the superfluous; who justify that conduct on the ground that it has the sanction of the cosmic process, and is the only way of ensuring the progress of the race ; who, if

---

[1] Whether the struggle for existence with the state of nature and with other societies, so far as they stand in the relation of the state of nature with it, exerts a selective influence upon modern society, and in what direction, are questions not easy to answer. The problem of the effect of military and industrial warfare upon those who wage it is very complicated.

they are consistent, must rank medicine among the black arts and count the physician a mischievous preserver of the unfit; on whose matrimonial undertakings the principles of the stud have the chief influence; whose whole lives, therefore, are an education in the noble art of suppressing natural affection and sympathy, are not likely to have any large stock of these commodities left. But, without them, there is no conscience, nor any restraint on the conduct of men, except the calculation of self-interest, the balancing of certain present gratifications against doubtful future pains; and experience tells us how much that is worth. Every day, we see firm believers in the hell of the theologians commit acts by which, as they believe when cool, they risk eternal punishment; while they hold back from those which are opposed to the sympathies of their associates.

XIII

That progressive modification of civilization which passes by the name of the "evolution of society," is, in fact, a process of an essentially different character, both from that which brings about the evolution of species, in the state of nature, and from that which gives rise to the evolution of varieties, in the state of art.

There can be no doubt that vast changes have taken place in English civilization since the reign of the Tudors. But I am not aware of a particle of evidence in favour of the conclusion that this evolutionary process has been accompanied by any modification of the physical, or the mental, characters of the men who have been the subjects of it. I have not met with any grounds for suspecting that the average Englishmen of to-day are sensibly different from those that Shakspere knew and drew. We look into his magic mirror of the Elizabethan age, and behold, nowise darkly, the presentment of ourselves.

During these three centuries, from the reign of Elizabeth to that of Victoria, the struggle for existence between man and man has been so largely restrained among the great mass of the population (except for one or two short intervals of civil war), that it can have had little, or no, selective operation. As to anything comparable to direct selection, it has been practised on so small a scale that it may also be neglected. The criminal law, in so far as by putting to death, or by subjecting to long periods of imprisonment, those who infringe its provisions, it prevents the propagation of hereditary criminal tendencies; and the poor-law, in so far as it separates married couples, whose destitution arises from hereditary defects of character, are doubtless selective agents operating in favour of the non-criminal and the more effective members of society. But the proportion of the population which they influence is very small; and, generally, the hereditary criminal and the hereditary pauper have propagated their kind before the law affects them. In a large proportion of cases, crime and pauperism have nothing to do with heredity; but are the consequence, partly, of circumstances and, partly, of the possession of qualities, which, under different conditions of life, might have excited esteem and even admiration. It was a shrewd man of the world who, in discussing sewage problems, remarked that dirt is riches in the wrong place; and that sound aphorism has moral applications. The benevolence and open-handed generosity which adorn a rich man, may make a pauper of a poor one; the energy and courage to which the successful soldier owes his rise, the cool and daring subtlety to which the great financier owes his fortune, may very easily, under unfavourable conditions, lead their possessors to the gallows, or to the hulks. Moreover, it is fairly probable that the children of a 'failure' will receive from their other parent just that little modification of character which makes all the difference. I sometimes wonder whether people, who talk so freely about extirpating the unfit, ever dispassionately consider their own history. Surely, one must be very 'fit,' indeed, not to know of an occasion, or perhaps two, in one's life, when it would have been only too easy to qualify for a place among the 'unfit.'

In my belief the innate qualities, physical, intellectual, and moral, of our nation have remained substantially the same for the last four or five centuries. If the struggle for existence has affected us to any serious extent (and I doubt it) it has been, indirectly, through our military and industrial wars with other nations.

XIV

What is often called the struggle for existence in society (I plead guilty to having used the term too loosely myself), is a contest, not for the means of existence, but for the means of enjoyment. Those who occupy the first places in this practical competitive examination are the rich and the influential; those who fail, more or less, occupy the lower places, down to the squalid obscurity of the pauper and the criminal. Upon the most liberal estimate, I suppose the former group will not amount to two per cent. of the population. I doubt if the latter exceeds another two per cent.; but let it be supposed, for the sake of argument, that it is as great as five per cent.[1]

As it is only in the latter group that anything comparable to the struggle for existence in the state of nature can take place; as it is only

---

[1] Those who read the last Essay in this volume will not accuse me of wishing to attenuate the evil of the existence of this group, whether great or small.

among this twentieth of the whole people that
numerous men, women, and children die of rapid
or slow starvation, or of the diseases incidental to
permanently bad conditions of life ; and as there
is nothing to prevent their multiplication before
they are killed off, while, in spite of greater
infant mortality, they increase faster than the
rich ; it seems clear that the struggle for existence in this class can have no appreciable selective influence upon the other 95 per cent. of
the population.

What sort of a sheep breeder would he be who
should content himself with picking out the worst
fifty out of a thousand, leaving them on a
barren common till the weakest starved, and then
letting the survivors go back to mix with the rest ?
And the parallel is too favourable ; since in a
large number of cases, the actual poor and the
convicted criminals are neither the weakest nor
the worst.

In the struggle for the means of enjoyment,
the qualities which ensure success are energy,
industry, intellectual capacity, tenacity of purpose,
and, at least as much sympathy as is necessary
to make a man understand the feelings of his
fellows. Were there none of those artificial arrangements by which fools and knaves are kept at
the top of society instead of sinking to their natural
place at the bottom,[1] the struggle for the means of

[1] I have elsewhere lamented the absence from society of

enjoyment would ensure a constant circulation of the human units of the social compound, from the bottom to the top and from the top to the bottom. The survivors of the contest, those who continued to form the great bulk of the polity, would not be those 'fittest' who got to the very top, but the great body of the moderately " fit," whose numbers and superior propagative power, enable them always to swamp the exceptionally endowed minority.

I think it must be obvious to every one, that, whether we consider the internal or the external interests of society, it is desirable they should be in the hands of those who are endowed with the largest share of energy, of industry, of intellectual capacity, of tenacity of purpose, while they are not devoid of sympathetic humanity; and, in so far as the struggle for the means of enjoyment tends to place such men in possession of wealth and influence, it is a process which tends to the good of society. But the process, as we have seen, has no real resemblance to that which adapts living beings to current conditions in the state of nature; nor any to the artificial selection of the horti-culturist.

a machinery for facilitating the descent of incapacity.   "Ad-ministrative Nihilism."   *Collected Essays*, vol. i. p. 54.

XV

To return, once more, to the parallel of horticulture. In the modern world, the gardening of men by themselves is practically restricted to the performance, not of selection, but of that other function of the gardener, the creation of conditions more favourable than those of the state of nature; to the end of facilitating the free expansion of the innate faculties of the citizen, so far as it is consistent with the general good. And the business of the moral and political philosopher appears to me to be the ascertainment, by the same method of observation, experiment, and ratiocination, as is practised in other kinds of scientific work, of the course of conduct which will best conduce to that end.

But, supposing this course of conduct to be scientifically determined and carefully followed out, it cannot put an end to the struggle for existence in the state of nature; and it will not so much as tend, in any way, to the adaptation of man to that state. Even should the whole human race be absorbed in one vast polity, within which "absolute political justice" reigns, the struggle for existence with the state of nature outside it, and the tendency to the return of the struggle within, in consequence of over-multiplication, will remain; and, unless men's inheritance from the ancestors who fought a good fight in the state of nature, their dose of original sin, is rooted out by some method at present unrevealed, at any rate to disbelievers in supernaturalism, every child born into the world will still bring with him the instinct of unlimited self-assertion. He will have to learn the lesson of self-restraint and renunciation. But the practice of self-restraint and renunciation is not happiness, though it may be something much better.

That man, as a 'political animal,' is susceptible of a vast amount of improvement, by education, by instruction, and by the application of his intelligence to the adaptation of the conditions of life to his higher needs, I entertain not the slightest doubt. But, so long as he remains liable to error, intellectual or moral; so long as he is compelled to be perpetually on guard against the cosmic forces, whose ends are not his ends, without and within himself; so long as he is haunted by inexpugnable memories and hopeless aspirations; so long as the recognition of his intellectual limitations forces him to acknowledge his incapacity to penetrate the mystery of existence; the prospect of attaining untroubled happiness, or of a state which can, even remotely, deserve the title of perfection, appears to me to be as misleading an illusion as ever was dangled before the eyes of poor humanity. And there have been many of them.

That which lies before the human race is a constant struggle to maintain and improve, in opposition to the State of Nature, the State of Art of an organized polity; in which, and by which, man may develop a worthy civilization, capable of maintaining and constantly improving itself, until the evolution of our globe shall have entered so far upon its downward course that the cosmic process resumes its sway; and, once more, the State of Nature prevails over the surface of our planet.

Note (see p. 30).—It seems the fashion nowadays to ignore Hartley; though, a century and a half ago, he not only laid the foundations but built up much of the superstructure of a true theory of the Evolution of the intellectual and moral faculties. He speaks of what I have termed the ethical process as "our Progress from Self-interest to Self-annihilation." *Observations on Man* (1749), vol. ii. p. 281.

II

EVOLUTION AND ETHICS

[*The Romanes Lecture*, 1893]

Soleo enim et in aliena castra transire, non tanquam transfuga sed tanquam explorator. (L. ANNÆI SENECÆ EPIST. II. 4.)

THERE is a delightful child's story, known by the title of " Jack and the Bean-stalk," with which my contemporaries who are present will be familiar. But so many of our grave and reverend juniors have been brought up on severer intellectual diet, and, perhaps, have become acquainted with fairyland only through primers of comparative mythology, that it may be needful to give an outline of the tale. It is a legend of a bean-plant, which grows and grows until it reaches the high heavens and there spreads out into a vast canopy of foliage. The hero, being moved to climb the stalk, discovers that the leafy expanse supports a world composed of the same elements as that below, but yet strangely new; and his adventures there, on which I may not dwell, must have com-

pletely changed his views of the nature of things;
though the story, not having been composed by,
or for, philosophers, has nothing to say about
views.

My present enterprise has a certain analogy to
that of the daring adventurer. I beg you to
accompany me in an attempt to reach a world
which, to many, is probably strange, by the help
of a bean. It is, as you know, a simple, inert-
looking thing. Yet, if planted under proper conditions, of which sufficient warmth is one of the
most important, it manifests active powers of a
very remarkable kind. A small green seedling
emerges, rises to the surface of the soil, rapidly
increases in size and, at the same time, undergoes
a series of metamorphoses which do not excite our
wonder as much as those which meet us in
legendary history, merely because they are to be
seen every day and all day long.

By insensible steps, the plant builds itself up
into a large and various fabric of root, stem, leaves,
flowers, and fruit, every one moulded within and
without in accordance with an extremely complex
but, at the same time, minutely defined pattern.
In each of these complicated structures, as in their
smallest constituents, there is an immanent energy
which, in harmony with that resident in all the
others, incessantly works towards the maintenance
of the whole and the efficient performance of the
part which it has to play in the economy of nature.

But no sooner has the edifice, reared with such
exact elaboration, attained completeness, than it
begins to crumble. By degrees, the plant withers
and disappears from view, leaving behind more or
fewer apparently inert and simple bodies, just like
the bean from which it sprang; and, like it, endowed with the potentiality of giving rise to a
similar cycle of manifestations.

Neither the poetic nor the scientific imagination
is put to much strain in the search after analogies
with this process of going forth and, as it were,
returning to the starting-point. It may be likened
to the ascent and descent of a slung stone, or
the course of an arrow along its trajectory. Or
we may say that the living energy takes first an
upward and then a downward road. Or it may
seem preferable to compare the expansion of the
germ into the full-grown plant, to the unfolding
of a fan, or to the rolling forth and widening of a
stream; and thus to arrive at the conception of
'development,' or 'evolution.' Here as elsewhere, names are 'noise and smoke'; the important point is to have a clear and adequate
conception of the fact signified by a name. And,
in this case, the fact is the Sisyphæan process, in
the course of which, the living and growing plant
passes from the relative simplicity and latent
potentiality of the seed to the full epiphany of a
highly differentiated type, thence to fall back to
simplicity and potentiality.

The value of a strong intellectual grasp of the nature of this process lies in the circumstance that what is true of the bean is true of living things in general. From very low forms up to the highest —in the animal no less than in the vegetable kingdom—the process of life presents the same appearance [1] of cyclical evolution. Nay, we have but to cast our eyes over the rest of the world and cyclical change presents itself on all sides. It meets us in the water that flows to the sea and returns to the springs; in the heavenly bodies that wax and wane, go and return to their places; in the inexorable sequence of the ages of man's life; in that successive rise, apogee, and fall of dynasties and of states which is the most prominent topic of civil history.

As no man fording a swift stream can dip his foot twice into the same water, so no man can, with exactness, affirm of anything in the sensible world that it is.[2] As he utters the words, nay, as he thinks them, the predicate ceases to be applicable; the present has become the past; the 'is' should be 'was.' And the more we learn of the nature of things, the more evident is it that what we call rest is only unperceived activity; that seeming peace is silent but strenuous battle. In every part, at every moment, the state of the cosmos is the expression of a transitory adjustment of contending forces; a scene of strife, in which all the combatants fall in turn. What is true of each part, is true of the whole. Natural knowledge tends more and more to the conclusion that " all the choir of heaven and furniture of the earth " are the transitory forms of parcels of cosmic substance wending along the road of evolution, from nebulous potentiality, through endless growths of sun and planet and satellite; through all varieties of matter; through infinite diversities of life and thought; possibly, through modes of being of which we neither have a conception, nor are competent to form any, back to the indefinable latency from which they arose. Thus the most obvious attribute of the cosmos is its impermanence. It assumes the aspect not so much of a permanent entity as of a changeful process, in which naught endures save the flow of energy and the rational order which pervades it.

We have climbed our bean-stalk and have reached a wonderland in which the common and the familiar become things new and strange. In the exploration of the cosmic process thus typified, the highest intelligence of man finds inexhaustible employment; giants are subdued to our service; and the spiritual affections of the contemplative philosopher are engaged by beauties worthy of eternal constancy.

But there is another aspect of the cosmic process, so perfect as a mechanism, so beautiful as a work of art. Where the cosmopoietic energy works through sentient beings, there arises, among its other manifestations, that which we call pain or suffering. This baleful product of evolution increases in quantity and in intensity, with advancing grades of animal organization, until it attains its highest level in man. Further, the consummation is not reached in man, the mere animal; nor in man, the whole or half savage; but only in man, the member of an organized polity. And it is a necessary consequence of his attempt to live in this way; that is, under those conditions which are essential to the full development of his noblest powers.

Man, the animal, in fact, has worked his way to the headship of the sentient world, and has become the superb animal which he is, in virtue of his success in the struggle for existence. The conditions having been of a certain order, man's organization has adjusted itself to them better than that of his competitors in the cosmic strife. In the case of mankind, the self-assertion, the unscrupulous seizing upon all that can be grasped, the tenacious holding of all that can be kept, which constitute the essence of the struggle for existence, have answered. For his successful progress, throughout the savage state, man has been largely indebted to those qualities which he shares with the ape and the tiger; his exceptional physical organization; his cunning, his sociability, his curiosity, and his imitativeness; his ruthless and ferocious destructiveness when his anger is roused by opposition.

But, in proportion as men have passed from anarchy to social organization, and in proportion as civilization has grown in worth, these deeply ingrained serviceable qualities have become defects. After the manner of successful persons, civilized man would gladly kick down the ladder by which he has climbed. He would be only too pleased to see 'the ape and tiger die.' But they decline to suit his convenience; and the unwelcome intrusion of these boon companions of his hot youth into the ranged existence of civil life adds pains and griefs, innumerable and immeasurably great, to those which the cosmic process necessarily brings on the mere animal. In fact, civilized man brands all these ape and tiger promptings with the name of sins; he punishes many of the acts which flow from them as crimes; and, in extreme cases, he does his best to put an end to the survival of the fittest of former days by axe and rope.

I have said that civilized man has reached this point; the assertion is perhaps too broad and general; I had better put it that ethical man has attained thereto. The science of ethics professes to furnish us with a reasoned rule of life; to tell us what is right action and why it is so. Whatever differences of opinion may exist among experts, there is a general consensus that the ape and tiger methods of the struggle for existence are not reconcilable with sound ethical principles.

The hero of our story descended the bean-stalk, and came back to the common world, where fare and work were alike hard ; where ugly competitors were much commoner than beautiful princesses ; and where the everlasting battle with self was much less sure to be crowned with victory than a turn-to with a giant. We have done the like. Thousands upon thousands of our fellows, thousands of years ago, have preceded us in finding themselves face to face with the same dread problem of evil. They also have seen that the cosmic process is evolution; that it is full of wonder, full of beauty, and, at the same time, full of pain. They have sought to discover the bearing of these great facts on ethics; to find out whether there is, or is not, a sanction for morality in the ways of the cosmos.

Theories of the universe, in which the conception of evolution plays a leading part, were extant at least six centuries before our era. Certain knowledge of them, in the fifth century, reaches us from localities as distant as the valley of the Ganges and the Asiatic coasts of the Ægean. To the early philosophers of Hindostan, no less than to those of Ionia, the salient and characteristic feature of the phenomenal world was its changefulness; the unresting flow of all things, through birth to visible being and thence to not being, in which they could discern no sign of a beginning and for which they saw no prospect of an ending. It was no less plain to some of these antique fore-runners of modern philosophy that suffering is the badge of all the tribe of sentient things; that it is no accidental accompaniment, but an essential constituent of the cosmic process. The energetic Greek might find fierce joys in a world in which 'strife is father and king'; but the old Aryan spirit was subdued to quietism in the Indian sage; the mist of suffering which spread over humanity hid everything else from his view; to him life was one with suffering and suffering with life.

In Hindostan, as in Ionia, a period of relatively high and tolerably stable civilization had succeeded long ages of semi-barbarism and struggle. Out of wealth and security had come leisure and refine-ment, and, close at their heels, had followed the malady of thought. To the struggle for bare existence, which never ends, though it may be alleviated and partially disguised for a fortunate few, succeeded the struggle to make existence intelligible and to bring the order of things into harmony with the moral sense of man, which also never ends, but, for the thinking few, becomes keener with every increase of knowledge and with every step towards the realization of a worthy ideal of life.

Two thousand five hundred years ago, the value of civilization was as apparent as it is now; then, as now, it was obvious that only in the garden of an orderly polity can the finest fruits humanity is capable of bearing be produced. But it had also become evident that the blessings of culture were not unmixed. The garden was apt to turn into a hothouse. The stimulation of the senses, the pampering of the emotions, endlessly multiplied the sources of pleasure. The constant widening of the intellectual field indefinitely extended the range of that especially human faculty of looking before and after, which adds to the fleeting present those old and new worlds of the past and the future, wherein men dwell the more the higher their culture. But that very sharpening of the sense and that subtle refinement of emotion, which brought such a wealth of pleasures, were fatally attended by a proportional enlargement of the capacity for suffering; and the divine faculty of imagination, while it created new heavens and new earths, provided them with the corresponding hells of futile regret for the past and morbid anxiety for the future.[3] Finally, the inevitable penalty of over-stimulation, exhaustion, opened the gates of civilization to its great enemy, ennui; the stale and flat weariness when man delights not, nor woman neither; when all things are vanity and vexation; and life seems not worth living except to escape the bore of dying.

Even purely intellectual progress brings about its revenges. Problems settled in a rough and ready way by rude men, absorbed in action, demand renewed attention and show themselves to be still unread riddles when men have time to think. The beneficent demon, doubt, whose name is Legion and who dwells amongst the tombs of old faiths, enters into mankind and thenceforth refuses to be cast out. Sacred customs, venerable dooms of ancestral wisdom, hallowed by tradition and professing to hold good for all time, are put to the question. Cultured reflection asks for their credentials; judges them by its own standards; finally, gathers those of which it approves into ethical systems, in which the reasoning is rarely much more than a decent pretext for the adoption of foregone conclusions.

One of the oldest and most important elements in such systems is the conception of justice. Society is impossible unless those who are associated agree to observe certain rules of conduct towards one another; its stability depends on the steadiness with which they abide by that agreement; and, so far as they waver, that mutual trust which is the bond of society is weakened or destroyed. Wolves could not hunt in packs except for the real, though unexpressed, understanding that they should not attack one another during the chase. The most rudimentary polity is a pack of men living under the like tacit, or expressed, understanding; and having made the very important advance upon wolf society, that they agree to use the force of the whole body against individuals who violate it and in favour of those who observe it. This observance of a common understanding, with the consequent distribution of punishments and rewards according to accepted rules, received the name of justice, while the contrary was called injustice. Early ethics did not take much note of the animus of the violator of the rules. But civilization could not advance far, without the establishment of a capital distinction between the case of involuntary and that of wilful misdeed; between a merely wrong action and a guilty one. And, with increasing refinement of moral appreciation, the problem of desert, which arises out of this distinction, acquired more and more theoretical and practical importance. If life must be given for life, yet it was recognized that the unintentional slayer did not altogether deserve death; and, by a sort of compromise between the public and the private conception of justice, a sanctuary was provided in which he might take refuge from the avenger of blood.

The idea of justice thus underwent a gradual sublimation from punishment and reward according to acts, to punishment and reward according to desert; or, in other words, according to motive. Righteousness, that is, action from right motive, not only became synonymous with justice, but the positive constituent of innocence and the very heart of goodness.

Now when the ancient sage, whether Indian or Greek, who had attained to this conception of goodness, looked the world, and especially human life, in the face, he found it as hard as we do to bring the course of evolution into harmony with even the elementary requirements of the ethical ideal of the just and the good.

If there is one thing plainer than another, it is that neither the pleasures nor the pains of life, in the merely animal world, are distributed according to desert; for it is admittedly impossible for the lower orders of sentient beings to deserve either the one or the other. If there is a generalization from the facts of human life which has the assent of thoughtful men in every age and country, it is that the violator of ethical rules constantly escapes the punishment which he deserves; that the wicked flourishes like a green bay tree, while the righteous begs his bread; that the sins of the fathers are visited upon the children; that, in the realm of nature, ignorance is punished just as severely as wilful wrong; and that thousands upon thousands of innocent beings suffer for the crime, or the unintentional trespass, of one.

Greek and Semite and Indian are agreed upon

this subject. The book of Job is at one with the
"Works and Days" and the Buddhist Sutras;
the Psalmist and the Preacher of Israel, with the
Tragic Poets of Greece. What is a more common
motive of the ancient tragedy in fact, than the
unfathomable injustice of the nature of things;
what is more deeply felt to be true than its pre-
sentation of the destruction of the blameless by
the work of his own hands, or by the fatal operation of the sins of others? Surely Œdipus was
pure of heart; it was the natural sequence of
events—the cosmic process—which drove him, in
all innocence, to slay his father and become the
husband of his mother, to the desolation of his
people and his own headlong ruin. Or to step, for
a moment, beyond the chronological limits I have
set myself, what constitutes the sempiternal attraction of Hamlet but the appeal to deepest
experience of that history of a no less blameless
dreamer, dragged, in spite of himself, into a world
out of joint; involved in a tangle of crime and
misery, created by one of the prime agents of the
cosmic process as it works in and through man?

Thus, brought before the tribunal of ethics, the
cosmos might well seem to stand condemned.
The conscience of man revolted against the moral
indifference of nature, and the microcosmic atom
should have found the illimitable macrocosm
guilty. But few, or none, ventured to record that
verdict.

In the great Semitic trial of this issue, Job takes refuge in silence and submission; the Indian and the Greek, less wise perhaps, attempt to reconcile the irreconcilable and plead for the defendant. To this end, the Greeks invented Theodicies; while the Indians devised what, in its ultimate form, must rather be termed a Cosmodicy. For, though Buddhism recognizes gods many and lords many, they are products of the cosmic process; and transitory, however long enduring, manifestations of its eternal activity. In the doctrine of transmigration, whatever its origin, Brahminical and Buddhist speculation found, ready to hand,[4] the means of constructing a plausible vindication of the ways of the cosmos to man. If this world is full of pain and sorrow; if grief and evil fall, like the rain, upon both the just and the unjust; it is because, like the rain, they are links in the endless chain of natural causation by which past, present, and future are indissolubly connected; and there is no more injustice in the one case than in the other. Every sentient being is reaping as it has sown; if not in this life, then in one or other of the infinite series of antecedent existences of which it is the latest term. The present distribution of good and evil is, therefore, the algebraical sum of accumulated positive and negative deserts; or, rather, it depends on the floating balance of the account. For it was not thought necessary that a complete

settlement should ever take place. Arrears might
stand over as a sort of 'hanging gale'; a period
of celestial happiness just earned might be succeeded by ages of torment in a hideous nether
world, the balance still overdue for some remote
ancestral error.[5]

Whether the cosmic process looks any more
moral than at first, after such a vindication, may
perhaps be questioned. Yet this plea of justification is not less plausible than others; and none
but very hasty thinkers will reject it on the
ground of inherent absurdity. Like the doctrine
of evolution itself, that of transmigration has its
roots in the world of reality; and it may claim
such support as the great argument from analogy
is capable of supplying.

Everyday experience familiarizes us with the
facts which are grouped under the name of heredity. Every one of us bears upon him obvious
marks of his parentage, perhaps of remoter relationships. More particularly, the sum of tendencies to act in a certain way, which we call
"character," is often to be traced through a long
series of progenitors and collaterals. So we may
justly say that this 'character'—this moral and
intellectual essence of a man—does veritably pass
over from one fleshly tabernacle to another, and
does really transmigrate from generation to
generation. In the new-born infant, the character
of the stock lies latent, and the Ego is little more

than a bundle of potentialities. But, very early, these become actualities; from childhood to age they manifest themselves in dulness or brightness, weakness or strength, viciousness or uprightness; and with each feature modified by confluence with another character, if by nothing else, the character passes on to its incarnation in new bodies.

The Indian philosophers called character, as thus defined, 'karma.'[6]   It is this karma which passed from life to life and linked them in the chain of transmigrations; and they held that it is modified in each life, not merely by confluence of parentage, but by its own acts. They were, in fact, strong believers in the theory, so much disputed just at present, of the hereditary transmission of acquired characters. That the manifestation of the tendencies of a character may be greatly facilitated, or impeded, by conditions, of which self-discipline, or the absence of it, are among the most important, is indubitable; but that the character itself is modified in this way is by no means so certain; it is not so sure that the transmitted character of an evil liver is worse, or that of a righteous man better, than that which he received. Indian philosophy, however, did not admit of any doubt on this subject; the belief in the influence of conditions, notably of self-discipline, on the karma was not merely a necessary postulate of its theory of retribution, but it presented the only way of escape from the endless round of transmigrations.

The earlier forms of Indian philosophy agreed with those prevalent in our own times, in supposing the existence of a permanent reality, or ' substance,' beneath the shifting series of phenomena, whether of matter or of mind. The substance of the cosmos was ' Brahma,' that of the individual man ' Atman '; and the latter was separated from the former only, if I may so speak, by its phenomenal envelope, by the casing of sensations, thoughts and desires, pleasures and pains, which make up the illusive phantasmagoria of life. This the ignorant take for reality; their ' Atman ' therefore remains eternally imprisoned in delusions, bound by the fetters of desire and scourged by the whip of misery. But the man who has attained enlightenment sees that the apparent reality is mere illusion, or, as was said a couple of thousand years later, that there is nothing good nor bad but thinking makes it so. If the cosmos " is just and of our pleasant vices makes instruments to scourge us," it would seem that the only way to escape from our heritage of evil is to destroy that fountain of desire whence our vices flow; to refuse any longer to be the instruments of the evolutionary process, and withdraw from the struggle for existence. If the karma is modifiable by self-discipline, if its coarser desires, one after another, can be extinguished, the ultimate fundamental desire of self-assertion, or the desire to be, may also be destroyed.[7] Then the bubble of illusion will burst, and the freed individual 'Atman' will lose itself in the universal 'Brahma.'

Such seems to have been the pre-Buddhistic conception of salvation, and of the way to be followed by those who would attain thereto. No more thorough mortification of the flesh has ever been attempted than that achieved by the Indian ascetic anchorite; no later monachism has so nearly succeeded in reducing the human mind to that condition of impassive quasi-somnambulism, which, but for its acknowledged holiness, might run the risk of being confounded with idiocy.

And this salvation, it will be observed, was to be attained through knowledge, and by action based on that knowledge; just as the experimenter, who would obtain a certain physical or chemical result, must have a knowledge of the natural laws involved and the persistent disciplined will adequate to carry out all the various operations required. The supernatural, in our sense of the term, was entirely excluded. There was no external power which could affect the sequence of cause and effect which gives rise to karma; none but the will of the subject of the karma which could put an end to it.

Only one rule of conduct could be based upon the remarkable theory of which I have endeavoured to give a reasoned outline. It was folly to continue to exist when an overplus of pain was certain; and the probabilities in favour of the increase of misery with the prolongation of existence, were so overwhelming. Slaying the body only made matters worse; there was nothing for it but to slay the soul by the voluntary arrest of all its activities. Property, social ties, family affections, common companionship, must be abandoned; the most natural appetites, even that for food, must be suppressed, or at least minimized; until all that remained of a man was the impassive, extenuated, mendicant monk, self-hypnotised into cataleptic trances, which the deluded mystic took for foretastes of the final union with Brahma.

The founder of Buddhism accepted the chief postulates demanded by his predecessors. But he was not satisfied with the practical annihilation involved in merging the individual existence in the unconditioned—the Atman in Brahma. It would seem that the admission of the existence of any substance whatever—even of the tenuity of that which has neither quality nor energy and of which no predicate whatever can be asserted— appeared to him to be a danger and a snare. Though reduced to a hypostatized negation, Brahma was not to be trusted; so long as entity was there, it might conceivably resume the weary round of evolution, with all its train of immeasurable miseries. Gautama got rid of even that shade of a shadow of permanent existence by a metaphysical *tour de force* of great interest to the student of philosophy, seeing that it supplies the wanting half of Bishop Berkeley's well-known idealistic argument.

Granting the premises, I am not aware of any escape from Berkeley's conclusion, that the 'substance' of matter is a metaphysical unknown quantity, of the existence of which there is no proof. What Berkeley does not seem to have so clearly perceived is that the non-existence of a substance of mind is equally arguable; and that the result of the impartial applications of his reasonings is the reduction of the All to co-existences and sequences of phenomena, beneath and beyond which there is nothing cognoscible. It is a remarkable indication of the subtlety of Indian speculation that Gautama should have seen deeper than the greatest of modern idealists; though it must be admitted that, if some of Berkeley's reasonings respecting the nature of spirit are pushed home, they reach pretty much the same conclusion.[8]

Accepting the prevalent Brahminical doctrine that the whole cosmos, celestial, terrestrial, and infernal, with its population of gods and other celestial beings, of sentient animals, of Mara and his devils, is incessantly shifting through recurring cycles of production and destruction, in each of which every human being has his transmigratory representative, Gautama proceeded to eliminate substance altogether; and to reduce the cosmos to a mere flow of sensations, emotions, volitions, and thoughts, devoid of any substratum. As, on the surface of a stream of water, we see ripples and whirlpools, which last for a while and then vanish with the causes that gave rise to them, so what seem individual existences are mere temporary associations of phenomena circling round a centre, "like a dog tied to a post." In the whole universe there is nothing permanent, no eternal substance either of mind or of matter. Personality is a metaphysical fancy; and in very truth, not only we, but all things, in the worlds without end of the cosmic phantasmagoria, are such stuff as dreams are made of.

What then becomes of karma? Karma remains untouched. As the peculiar form of energy we call magnetism may be transmitted from a loadstone to a piece of steel, from the steel to a piece of nickel, as it may be strengthened or weakened by the conditions to which it is subjected while resident in each piece, so it seems to have been conceived that karma might be transmitted from one phenomenal association to another by a sort of induction. However this may be, Gautama doubtless had a better guarantee for the abolition of transmigration, when no wrack of substance, either of Atman or of Brahma, was left behind when, in short, a man had but to dream that he willed not to dream, to put an end to all dreaming.

This end of life's dream is Nirvana. What Nirvana is the learned do not agree. But, since the best original authorities tell us there is neither desire nor activity, nor any possibility of phenomenal reappearance for the sage who has entered Nirvana, it may be safely said of this acme of Buddhistic philosophy—" the rest is silence."[9]

Thus there is no very great practical disagreement between Gautama and his predecessors with respect to the end of action; but it is otherwise as regards the means to that end. With just insight into human nature, Gautama declared extreme ascetic practices to be useless and indeed harmful. The appetites and the passions are not to be abolished by mere mortification of the body; they must, in addition, be attacked on their own ground and conquered by steady cultivation of the mental habits which oppose them; by universal benevolence; by the return of good for evil; by humility; by abstinence from evil thought; in short, by total renunciation of that self-assertion which is the essence of the cosmic process.

Doubtless, it is to these ethical qualities that Buddhism owes its marvellous success.[10] A system which knows no God in the western sense; which denies a soul to man; which counts the belief in immortality a blunder and the hope of it a sin;

which refuses any efficacy to prayer and sacrifice; which bids men look to nothing but their own efforts for salvation; which, in its original purity, knew nothing of vows of obedience, abhorred intolerance, and never sought the aid of the secular arm; yet spread over a considerable moiety of the Old World with marvellous rapidity, and is still, with whatever base admixture of foreign superstitions, the dominant creed of a large fraction of mankind.

Let us now set our faces westwards, towards Asia Minor and Greece and Italy, to view the rise and progress of another philosophy, apparently independent, but no less pervaded by the conception of evolution.[11]

The sages of Miletus were pronounced evolutionists; and, however dark may be some of the sayings of Heracleitus of Ephesus, who was probably a contemporary of Gautama, no better expressions of the essence of the modern doctrine of evolution can be found than are presented by some of his pithy aphorisms and striking metaphors.[12] Indeed, many of my present auditors must have observed that, more than once, I have borrowed from him in the brief exposition of the theory of evolution with which this discourse commenced.

But when the focus of Greek intellectual activity shifted to Athens, the leading minds concentrated their attention upon ethical problems. Forsaking the study of the macrocosm for that of the microcosm, they lost the key to the thought of the great Ephesian, which, I imagine, is more intelligible to us than it was to Socrates, or to Plato. Socrates, more especially, set the fashion of a kind of inverse agnosticism, by teaching that the problems of physics lie beyond the reach of the human intellect; that the attempt to solve them is essentially vain; that the one worthy object of investigation is the problem of ethical life; and his example was followed by the Cynics and the later Stoics. Even the comprehensive knowledge and the penetrating intellect of Aristotle failed to suggest to him that in holding the eternity of the world, within its present range of mutation, he was making a retrogressive step. The scientific heritage of Heracleitus passed into the hands neither of Plato nor of Aristotle, but into those of Democritus. But the world was not yet ready to receive the great conceptions of the philosopher of Abdera. It was reserved for the Stoics to return to the track marked out by the earlier philosophers; and, professing themselves disciples of Heracleitus, to develop the idea of evolution systematically. In doing this, they not only omitted some characteristic features of their master's teaching, but they made additions altogether foreign to it. One of the most influential of these importations was the transcen-

dental theism which had come into vogue. The
restless, fiery energy, operating according to law,
out of which all things emerge and into which
they return, in the endless successive cycles of
the great year; which creates and destroys worlds
as a wanton child builds up, and anon levels, sand
castles on the seashore; was metamorphosed into
a material world-soul and decked out with all the
attributes of ideal Divinity; not merely with infinite power and transcendent wisdom, but with
absolute goodness.

The consequences of this step were momentous.
For if the cosmos is the effect of an immanent,
omnipotent, and infinitely beneficent cause, the existence in it of real evil, still less of necessarily
inherent evil, is plainly inadmissible.[13] Yet the
universal experience of mankind testified then, as
now, that, whether we look within us or without
us, evil stares us in the face on all sides; that if
anything is real, pain and sorrow and wrong are
realities.

It would be a new thing in history if *à priori*
philosophers were daunted by the factious opposition of experience; and the Stoics were the last
men to allow themselves to be beaten by mere
facts. ' Give me a doctrine and I will find the
reasons for it,' said Chrysippus. So they perfected, if they did not invent, that ingenious and
plausible form of pleading, the Theodicy; for the
purpose of showing firstly, that there is no such

thing as evil; secondly, that if there is, it is the necessary correlate of good; and, moreover, that it is either due to our own fault, or inflicted for our benefit. Theodicies have been very popular in their time, and I believe that a numerous, though somewhat dwarfed, progeny of them still survives. So far as I know, they are all variations of the theme set forth in those famous six lines of the " Essay on Man," in which Pope sums up Boling-broke's reminiscences of stoical and other specu-lations of this kind—

> "All nature is but art, unknown to thee;
> All chance, direction which thou canst not see;
> All discord, harmony not understood;
> All partial evil, universal good;
> And spite of pride, in erring reason's spite
> One truth is clear : whatever is is right."

Yet, surely, if there are few more important truths than those enunciated in the first triad, the second is open to very grave objections. That there is a 'soul of good in things evil' is un-questionable; nor will any wise man deny the disciplinary value of pain and sorrow. But these considerations do not help us to see why the im-mense multitude of irresponsible sentient beings, which cannot profit by such discipline, should suffer; nor why, among the endless possibilities open to omnipotence—that of sinless, happy exist-ence among the rest—the actuality in which sin and misery abound should be that selected.

Surely it is mere cheap rhetoric to call arguments which have never yet been answered by even the meekest and the least rational of Optimists, suggestions of the pride of reason. As to the concluding aphorism, its fittest place would be as an inscription in letters of mud over the portal of some ' stye of Epicurus ';[14] for that is where the logical application of it to practice would land men, with every aspiration stifled and every effort paralyzed. Why try to set right what is right already ? Why strive to improve the best of all possible worlds ? Let us eat and drink, for as to-day all is right, so to-morrow all will be.

But the attempt of the Stoics to blind themselves to the reality of evil, as a necessary concomitant of the cosmic process, had less success than that of the Indian philosophers to exclude the reality of good from their purview. Unfortunately, it is much easier to shut one's eyes to good than to evil. Pain and sorrow knock at our doors more loudly than pleasure and happiness ; and the prints of their heavy footsteps are less easily effaced. Before the grim realities of practical life the pleasant fictions of optimism vanished. If this were the best of all possible worlds, it nevertheless proved itself a very inconvenient habitation for the ideal sage.

The stoical summary of the whole duty of man, ' Live according to nature,' would seem to imply that the cosmic process is an exemplar for human

conduct. Ethics would thus become applied
Natural History. In fact, a confused employment
of the maxim, in this sense, has done immeasurable mischief in later times. It has furnished an
axiomatic foundation for the philosophy of philo-
sophasters and for the moralizing of sentimentalists.
But the Stoics were, at bottom, not merely noble,
but sane, men; and if we look closely into what
they really meant by this ill-used phrase, it will
be found to present no justification for the mischievous conclusions that have been deduced
from it.

In the language of the Stoa, ' Nature ' was a
word of many meanings. There was the ' Nature '
of the cosmos and the ' Nature ' of man. In the
latter, the animal 'nature,' which man shares
with a moiety of the living part of the cosmos, was
distinguished from a higher 'nature.' Even in
this higher nature there were grades of rank.
The logical faculty is an instrument which may be
turned to account for any purpose. The passions
and the emotions are so closely tied to the lower
nature that they may be considered to be pathological, rather than normal, phenomena. The one
supreme, hegemonic, faculty, which constitutes the
essential ' nature ' of man, is most nearly represented by that which, in the language of a later
philosophy, has been called the pure reason. It is
this ' nature ' which holds up the ideal of the
supreme good and demands absolute submission of

the will to its behests. It is this which commands all men to love one another, to return good for evil, to regard one another as fellow-citizens of one great state. Indeed, seeing that the progress towards perfection of a civilized state, or polity, depends on the obedience of its members to these commands, the Stoics sometimes termed the pure reason the 'political' nature. Unfortunately, the sense of the adjective has undergone so much modification, that the application of it to that which commands the sacrifice of self to the common good would now sound almost grotesque.[15]

But what part is played by the theory of evolution in this view of ethics ? So far as I can discern, the ethical system of the Stoics, which is essentially intuitive, and reverences the categorical imperative as strongly as that of any later moralists, might have been just what it was if they had held any other theory ; whether that of special creation, on the one side, or that of the eternal existence of the present order, on the other.[16] To the Stoic, the cosmos had no importance for the conscience, except in so far as he chose to think it a pedagogue to virtue. The pertinacious optimism of our philosophers hid from them the actual state of the case. It prevented them from seeing that cosmic nature is no school of virtue, but the headquarters of the enemy of ethical nature. The logic of facts was necessary to convince them that the cosmos works through the lower nature of man, not for righteousness, but against it. And it finally drove them to confess that the existence of their ideal " wise man " was incompatible with the nature of things ; that even a passable approximation to that ideal was to be attained only at the cost of renunciation of the world and mortification, not merely of the flesh, but of all human affections. The state of perfection was that ' apatheia '[17] in which desire, though it may still be felt, is powerless to move the will, reduced to the sole function of executing the commands of pure reason. Even this residuum of activity was to be regarded as a temporary loan, as an efflux of the divine world-pervading spirit, chafing at its imprisonment in the flesh, until such time as death enabled it to return to its source in the all-pervading logos.

I find it difficult to discover any very great difference between Apatheia and Nirvana, except that stoical speculation agrees with pre-Buddhistic philosophy, rather than with the teachings of Gautama, in so far as it postulates a permanent substance equivalent to ' Brahma ' and ' Atman '; and that, in stoical practice, the adoption of the life of the mendicant cynic was held to be more a counsel of perfection than an indispensable condition of the higher life.

Thus the extremes touch. Greek thought and

Indian thought set out from ground common to both, diverge widely, develop under very different physical and moral conditions, and finally converge to practically the same end.

The Vedas and the Homeric epos set before us a world of rich and vigorous life, full of joyous fighting men

> That ever with a frolic welcome took
> The thunder and the sunshine . . . .

and who were ready to brave the very Gods themselves when their blood was up. A few centuries pass away, and under the influence of civilization the descendants of these men are ' sicklied o'er with the pale cast of thought '—frank pessimists, or, at best, make-believe optimists. The courage of the warlike stock may be as hardly tried as before, perhaps more hardly, but the enemy is self. The hero has become a monk. The man of action is replaced by the quietist, whose highest aspiration is to be the passive instrument of the divine Reason. By the Tiber, as by the Ganges, ethical man admits that the cosmos is too strong for him; and, destroying every bond which ties him to it by ascetic discipline, he seeks salvation in absolute renunciation.[18]

Modern thought is making a fresh start from the base whence Indian and Greek philosophy set out; and, the human mind being very much what it was six-and-twenty centuries ago, there is no ground for wonder if it presents indications of a tendency to move along the old lines to the same results.

We are more than sufficiently familiar with modern pessimism, at least as a speculation; for I cannot call to mind that any of its present votaries have sealed their faith by assuming the rags and the bowl of the mendicant Bhikku, or the cloak and the wallet of the Cynic. The obstacles placed in the way of sturdy vagrancy by an unphiloso-phical police have, perhaps, proved too formidable for philosophical consistency. We also know modern speculative optimism, with its perfectibility of the species, reign of peace, and lion and lamb transformation scenes; but one does not hear so much of it as one did forty years ago; indeed, I imagine it is to be met with more commonly at the tables of the healthy and wealthy, than in the congregations of the wise. The majority of us, I apprehend, profess neither pessimism nor optimism. We hold that the world is neither so good, nor so bad, as it conceivably might be; and, as most of us have reason, now and again, to discover that it can be. Those who have failed to experience the joys that make life worth living are, probably, in as small a minority as those who have never known the griefs that rob existence of its savour and turn its richest fruits into mere dust and ashes.

Further, I think I do not err in assuming that, however diverse their views on philosophical and religious matters, most men are agreed that the proportion of good and evil in life may be very sensibly affected by human action. I never heard anybody doubt that the evil may be thus increased, or diminished; and it would seem to follow that good must be similarly susceptible of addition or subtraction. Finally, to my knowledge, nobody professes to doubt that, so far forth as we possess a power of bettering things, it is our paramount duty to use it and to train all our intellect and energy to this supreme service of our kind.

Hence the pressing interest of the question, to what extent modern progress in natural knowledge, and, more especially, the general outcome of that progress in the doctrine of evolution, is competent to help us in the great work of helping one another?

The propounders of what are called the " ethics of evolution," when the ' evolution of ethics ' would usually better express the object of their speculations, adduce a number of more or less interesting facts and more or less sound arguments, in favour of the origin of the moral sentiments, in the same way as other natural phenomena, by a process of evolution. I have little doubt, for my own part, that they are on the right track; but as the immoral sentiments have no less been evolved, there is, so far, as much natural sanction for the

one as the other. The thief and the murderer
follow nature just as much as the philanthropist.
Cosmic evolution may teach us how the good and
the evil tendencies of man may have come about;
but, in itself, it is incompetent to furnish any
better reason why what we call good is preferable
to what we call evil than we had before. Some
day, I doubt not, we shall arrive at an understanding of the evolution of the æsthetic faculty; but
all the understanding in the world will neither
increase nor diminish the force of the intuition
that this is beautiful and that is ugly.

There is another fallacy which appears to me to
pervade the so-called " ethics of evolution." It is
the notion that because, on the whole, animals
and plants have advanced in perfection of organization by means of the struggle for existence and
the consequent 'survival of the fittest'; therefore
men in society, men as ethical beings, must look
to the same process to help them towards perfection. I suspect that this fallacy has arisen out
of the unfortunate ambiguity of the phrase 'survival of the fittest.' 'Fittest' has a connotation of
'best'; and about 'best' there hangs a moral
flavour. In cosmic nature, however, what is
'fittest' depends upon the conditions. Long since,[19]
I ventured to point out that if our hemisphere
were to cool again, the survival of the fittest might
bring about, in the vegetable kingdom, a population of more and more stunted and humbler and

humbler organisms, until the 'fittest' that survived might be nothing but lichens, diatoms, and
such microscopic organisms as those which give
red snow its colour; while, if it became hotter, the
pleasant valleys of the Thames and Isis might be
uninhabitable by any animated beings save those
that flourish in a tropical jungle. They, as the
fittest, the best adapted to the changed conditions,
would survive.

Men in society are undoubtedly subject to the
cosmic process. As among other animals, multiplication goes on without cessation, and involves
severe competition for the means of support. The
struggle for existence tends to eliminate those less
fitted to adapt themselves to the circumstances
of their existence. The strongest, the most self-
assertive, tend to tread down the weaker. But
the influence of the cosmic process on the evolution of society is the greater the more rudimentary
its civilization. Social progress means a checking
of the cosmic process at every step and the substitution for it of another, which may be called
the ethical process; the end of which is not the
survival of those who may happen to be the
fittest, in respect of the whole of the conditions
which obtain, but of those who are ethically the
best.[20]

As I have already urged, the practice of that
which is ethically best—what we call goodness or
virtue—involves a course of conduct which, in all

respects, is opposed to that which leads to success in the cosmic struggle for existence. In place of ruthless self-assertion it demands self-restraint; in place of thrusting aside, or treading down, all competitors, it requires that the individual shall not merely respect, but shall help his fellows; its influence is directed, not so much to the survival of the fittest, as to the fitting of as many as possible to survive. It repudiates the gladiatorial theory of existence. It demands that each man who enters into the enjoyment of the advantages of a polity shall be mindful of his debt to those who have laboriously constructed it; and shall take heed that no act of his weakens the fabric in which he has been permitted to live. Laws and moral precepts are directed to the end of curbing the cosmic process and reminding the individual of his duty to the community, to the protection and influence of which he owes, if not existence itself, at least the life of something better than a brutal savage.

It is from neglect of these plain considerations that the fanatical individualism [21] of our time attempts to apply the analogy of cosmic nature to society. Once more we have a misapplication of the stoical injunction to follow nature; the duties of the individual to the state are forgotten, and his tendencies to self-assertion are dignified by the name of rights. It is seriously debated whether the members of a community are justified in using their combined strength to constrain one of their number to contribute his share to the maintenance of it; or even to prevent him from doing his best to destroy it. The struggle for existence, which has done such admirable work in cosmic nature, must, it appears, be equally beneficent in the ethical sphere. Yet if that which I have insisted upon is true; if the cosmic process has no sort of relation to moral ends; if the imitation of it by man is inconsistent with the first principles of ethics; what becomes of this surprising theory?

Let us understand, once for all, that the ethical progress of society depends, not on imitating the cosmic process, still less in running away from it, but in combating it. It may seem an audacious proposal thus to pit the microcosm against the macrocosm and to set man to subdue nature to his higher ends; but I venture to think that the great intellectual difference between the ancient times with which we have been occupied and our day, lies in the solid foundation we have acquired for the hope that such an enterprise may meet with a certain measure of success.

The history of civilization details the steps by which men have succeeded in building up an artificial world within the cosmos. Fragile reed as he may be, man, as Pascal says, is a thinking reed:[22] there lies within him a fund of energy, operating intelligently and so far akin to that which pervades the universe, that it is competent to influence and modify the cosmic process. In virtue of his intelligence, the dwarf bends the Titan to his will. In every family, in every polity that has been established, the cosmic process in man has been restrained and otherwise modified by law and custom ; in surrounding nature, it has been similarly influenced by the art of the shepherd, the agriculturist, the artisan. As civilization has advanced, so has the extent of this interference increased ; until the organized and highly developed sciences and arts of the present day have endowed man with a command over the course of non-human nature greater than that once attributed to the magicians. The most impressive, I might say startling, of these changes have been brought about in the course of the last two centuries ; while a right comprehension of the process of life and of the means of influencing its manifestations is only just dawning upon us. We do not yet see our way beyond generalities ; and we are befogged by the obtrusion of false analogies and crude anticipations. But Astronomy, Physics, Chemistry, have all had to pass through similar phases, before they reached the stage at which their influence became an important factor in human affairs. Physiology, Psychology, Ethics, Political Science, must submit to the same ordeal. Yet it seems to me irrational to doubt that, at no distant period, they will work as great a revolution in the sphere of practice.

The theory of evolution encourages no millennial anticipations. If, for millions of years, our globe has taken the upward road, yet, some time, the summit will be reached and the downward route will be commenced. The most daring imagination will hardly venture upon the suggestion that the power and the intelligence of man can ever arrest the procession of the great year.

Moreover, the cosmic nature born with us and, to a large extent, necessary for our maintenance, is the outcome of millions of years of severe training, and it would be folly to imagine that a few centuries will suffice to subdue its masterfulness to purely ethical ends. Ethical nature may count upon having to reckon with a tenacious and powerful enemy as long as the world lasts. But, on the other hand, I see no limit to the extent to which intelligence and will, guided by sound principles of investigation, and organized in common effort, may modify the conditions of existence, for a period longer than that now covered by history. And much may be done to change the nature of man himself.[23] The intelligence which has converted the brother of the wolf into the faithful guardian of the flock ought to be able to do something towards curbing the instincts of savagery in civilized men.

But if we may permit ourselves a larger hope of abatement of the essential evil of the world than was possible to those who, in the infancy of exact knowledge, faced the problem of existence more than a score of centuries ago, I deem it an essential condition of the realization of that hope that we should cast aside the notion that the escape from pain and sorrow is the proper object of life.

We have long since emerged from the heroic childhood of our race, when good and evil could be met with the same ' frolic welcome '; the attempts to escape from evil, whether Indian or Greek, have ended in flight from the battle-field; it remains to us to throw aside the youthful over-confidence and the no less youthful discouragement of nonage. We are grown men, and must play the man

strong in will
To strive, to seek, to find, and not to yield,

cherishing the good that falls in our way, and bearing the evil, in and around us, with stout hearts set on diminishing it. So far, we all may strive in one faith towards one hope :

It may be that the gulfs will wash us down,
It may be we shall touch the Happy Isles,

. . . . but something ere the end,
Some work of noble note may yet be done. (²⁴)

NOTES

Note 1 (p. 49).

I HAVE been careful to speak of the "appearance" of cyclical evolution presented by living things; for, on critical examination, it will be found that the course of vegetable and of animal life is not exactly represented by the figure of a cycle which returns into itself. What actually happens, in all but the lowest organisms, is that one part of the growing germ (A) gives rise to tissues and organs; while another part (B) remains in its primitive condition, or is but slightly modified. The moiety A becomes the body of the adult and, sooner or later, perishes, while portions of the moiety B are detached and, as offspring, continue the life of the species. Thus, if we trace back an organism along the direct line of descent from its remotest ancestor, B, as a whole, has never suffered death; portions of it, only, have been cast off and died in each individual offspring.

Everybody is familiar with the way in which the "suckers" of a strawberry plant behave. A thin cylinder of living tissue keeps on growing at its free end, until it attains a considerable length. At successive intervals, it develops buds which grow into strawberry plants ; and these become independent by the death of the parts of the sucker which connect them.   The rest of the sucker, however, may go on living and growing indefinitely, and, circumstances remaining favourable, there is no obvious reason why it should ever die.   The living substance B, in a manner, answers to the sucker.   If we could restore the continuity which was once possessed by the portions of B, contained in all the individuals of a direct line of descent, they would form a sucker, or *stolon*, on which these individuals would be strung, and which would never have wholly died.

A species remains unchanged so long as the potentiality of development resident in B remains unaltered ; so long, *e.g.*, as the buds of the strawberry sucker tend to become typical strawberry plants.   In the case of the progressive evolution of a species, the developmental potentiality of B becomes of a higher and higher order.   In retrogressive evolution, the contrary would be the case.   The phenomena of atavism seem to show that retrogressive evolution, that is, the return of a species to one or other of its earlier forms, is a possibility to be reckoned with. The simplification of structure, which is so common in the parasitic members of a group, however, does not properly come under this head.   The worm-like, limbless *Lernœa* has no resemblance to any of the stages of development of the many-limbed active animals of the group to which it belongs.

Note 2 (p. 49)

Heracleitus says, Ποταμῷ γὰρ οὐκ ἔστι δὶς ἐμβῆναι τῷ αὐτῷ; but, to be strictly accurate, the river remains, though the water of which it is composed changes—just as a man retains his identity though the whole substance of his body is constantly shifting.   *

This is put very well by Seneca (Ep. lvii. i. 20, Ed. Ruhkopf): " Corpora nostra rapiuntur fluminum more, quidquid vides currit cum tempore; nihil ex his quæ videmus manet. Ego ipse dum loquor mutari ista, mutatus sum. Hoc est quod ait Heraclitus ' In idem flumen bis non descendimus.' Manet idem fluminis nomen, aqua transmissa est. Hoc in amne manifestius est quam in homine, sed nos quoque non minus velox cursus prætervehit."   †

Note 3 (p. 55).

" Multa bona nostra nobis nocent, timoris enim tormentum memoria reducit, providentia anticipat. Nemo tantum præsentibus miser est." (Seneca, Ed. v. 7.)   ‡

Among the many wise and weighty aphorisms of the Roman Bacon, few sound the realities of life more deeply than " Multa bona nostra nobis nocent." If there is a soul of good in things evil, it is at least equally true that there is a soul of evil in things good : for things, like men, have " les défauts de leurs qualités." It is one of the last lessons one learns from experience, but not the least important, that a heavy tax is levied upon all forms of success ; and that failure is one of the commonest disguises assumed by blessings.

Note 4 (p. 60).

" There is within the body of every man a soul which, at the death of the body, flies away from it like a bird out of a cage, and enters upon a new life . . . either in one of the heavens or one of the hells or on this earth. The only exception is the rare case of a man having in this life acquired a true knowledge of God. According to the pre-Buddhistic theory, the soul of such a man goes along the path of the Gods to God, and, being united with Him, enters upon an immortal life in which his individuality is not extinguished. In the latter theory, his soul is directly absorbed into the Great Soul, is lost in it, and has no longer any independent existence. The souls of all other men enter, after the death of the body, upon a new existence in one or other of the many different modes of being. If in heaven or hell, the soul itself becomes a god or demon without entering a body ; all superhuman beings, save the great gods, being looked upon as not eternal, but merely temporary creatures. If the soul returns to earth it may or may not enter a new body ; and this either of a human being, an animal, a plant, or even a material object. For all these are possessed of souls, and there is no essential difference between these souls and the souls of men—all being alike mere sparks of the Great Spirit, who is the only real existence." (Rhys Davids, *Hibbert Lectures*, 1881, p. 83.)

For what I have said about Indian Philosophy, I am particularly indebted to the luminous exposition of primitive Buddhism and its relations to earlier Hindu thought, which is given by Prof. Rhys Davids in his remarkable *Hibbert Lectures* for 1881, and *Buddhism* (1890). The only apology I can offer for the freedom with which I have borrowed from him in these notes, is my desire to leave no doubt as to my indebtedness. I have also found Dr. Oldenberg's *Buddha* (Ed. 2, 1890) very helpful. The origin of the theory of transmigration stated in the above extract is an unsolved problem. That it differs widely from the Egyptian metempsychosis is clear. In fact, since men usually people the other world with phantoms of this, the Egyptian doctrine would seem to presuppose the Indian as a more archaic belief.

Prof. Rhys Davids has fully insisted upon the ethical importance of the transmigration theory. " One of the latest speculations now being put forward among ourselves would seek to explain each man's character, and even his outward condition in life, by the character he inherited from his ancestors, a character gradually formed during a practically endless series of past existences, modified only by the conditions into which he was born, those very conditions being also, in like manner, the last result of a practically endless series of past causes. Gotama's speculation might be stated in the same words. But it attempted also to explain, in a way different from

that which would be adopted by the exponents of
the modern theory, that strange problem which it
is also the motive of the wonderful drama of the
book of Job to explain—the fact that the actual
distribution here of good fortune, or misery, is
entirely independent of the moral qualities which
men call good or bad.   We cannot wonder that a
teacher, whose whole system was so essentially an
ethical reformation, should have felt it incumbent
upon him to seek an explanation of this apparent
injustice.   And all the more so, since the belief he
had inherited, the theory of the transmigration of
souls, had provided a solution perfectly sufficient to
any one who could accept that belief."   (*Hibbert
Lectures*, p. 93.)   I should venture to suggest the
substitution of 'largely' for 'entirely' in the fore-
going passage.   Whether a ship makes a good or a
bad voyage is largely independent of the conduct of
the captain, but it is largely affected by that conduct.   Though powerless before a hurricane he may
weather many a bad gale.

Note 5 (p. 61).

The outward condition of the soul is, in each new
birth, determined by its actions in a previous birth ;
but by each action in succession, and not by the
balance struck after the evil has been reckoned off
against the good.   A good man who has once uttered
a slander may spend a hundred thousand years as a
god, in consequence of his goodness, and when the
power of his good actions is exhausted, may be born

as a dumb man on account of his transgression; and a robber who has once done an act of mercy, may come to life in a king's body as the result of his virtue, and then suffer torments for ages in hell or as a ghost without a body, or be re-born many times as a slave or an outcast, in consequence of his evil life.

"There is no escape, according to this theory, from the result of any act; though it is only the consequences of its own acts that each soul has to endure. The force has been set in motion by itself and can never stop; and its effect can never be foretold. If evil, it can never be modified or prevented, for it depends on a cause already completed, that is now for ever beyond the soul's control. There is even no continuing consciousness, no memory of the past that could guide the soul to any knowledge of its fate. The only advantage open to it is to add in this life to the sum of its good actions, that it may bear fruit with the rest. And even this can only happen in some future life under essentially the same conditions as the present one: subject, like the present one, to old age, decay, and death; and affording opportunity, like the present one, for the commission of errors, ignorances, or sins, which in their turn must inevitably produce their due effect of sickness, disability, or woe. Thus is the soul tossed about from life to life, from billow to billow in the great ocean of transmigration. And there is no escape save for the very few, who, during their birth as men, attain to a right knowledge of the Great Spirit: and thus enter into immortality, or, as the later philosophers taught, are absorbed into the

Divine Essence." (Rhys Davids, *Hibbert Lectures*, pp. 85, 86.)

The state after death thus imagined by the Hindu philosophers has a certain analogy to the purgatory of the Roman Church; except that escape from it is dependent, not on a divine decree modified, it may be, by sacerdotal or saintly intercession, but by the acts of the individual himself; and that while ultimate emergence into heavenly bliss of the good, or well-prayed for, Catholic is professedly assured, the chances in favour of the attainment of absorption, or of Nirvana, by any individual Hindu are extremely small.

Note 6 (p. 62).

" That part of the then prevalent transmigration theory which could not be proved false seemed to meet a deeply felt necessity, seemed to supply a moral cause which would explain the unequal distri-bution here of happinesss or woe, so utterly inconsis-tent with the present characters of men." Gautama " still therefore talked of men's previous existence, but by no means in the way that he is generally represented to have done." What he taught was "the transmigration of character." He held that after the death of any being, whether human or not, there survived nothing at all but that being's ' Karma,' the result, that is, of its mental and bodily actions. Every individual, whether human or divine, was the last inheritor and the last result of the Karma of a long series of past individuals—a series so long that its beginning is beyond the reach of calculation, and its end will be coincident with the destruction of the world." (Rhys Davids, *Hibbert Lectures*, p. 92.)

In the theory of evolution, the tendency of a germ to develop according to a certain specific type, *e.g.* of the kidney bean seed to grow into a plant having all the characters of *Phaseolus vulgaris*, is its ' Karma.' It is the " last inheritor and the last result " of all the conditions that have affected a line of ancestry which goes back for many millions of years to the time when life first appeared on the earth. The moiety B of the substance of the bean plant (see *Note* 1) is the last link in a once continuous chain extending from the primitive living substance : and the characters of the successive species to which it has given rise are the manifestations of its gradually modified Karma. As Prof. Rhys Davids aptly says, the snowdrop " is a snowdrop and not an oak, and just that kind of snowdrop, because it is the outcome of the Karma of an endless series of past existences." (*Hibbert Lectures*, p. 114.)

Note 7 (p. 64).

" It is interesting to notice that the very point which is the weakness of the theory—the supposed concentration of the effect of the Karma in one new being—presented itself to the early Buddhists themselves as a difficulty. They avoided it, partly by explaining that it was a particular thirst in the creature dying (a craving, Tanhā, which plays otherwise a great part in the Buddhist theory) which actually caused the birth of the new individual who was to inherit the Karma of the former one. But, how this took place, how the craving desire produced this effect, was acknowledged to be a mystery patent only to a Buddha." (Rhys Davids, *Hibbert Lectures*, p. 95.)

Among the many parallelisms of Stoicism and Buddhism, it is curious to find one for this Tanhā, 'thirst,' or 'craving desire' for life. Seneca writes (Epist. lxxvi. 18): "Si enim ullum aliud est bonum quam honestum, sequetur nos *aviditas vitæ* aviditas rerum vitam instruentium: quod est intolerabile infinitum, vagum."

Note 8 (p. 66).

"The distinguishing characteristic of Buddhism was that it started a new line, that it looked upon the deepest questions men have to solve from an entirely different standpoint. It swept away from the field of its vision the whole of the great soul-theory which had hitherto so completely filled and dominated the minds of the superstitious and the thoughtful alike. For the first time in the history of the world, it proclaimed a salvation which each man could gain for himself and by himself, in this world, during this life, without any the least reference to God, or to Gods, either great or small. Like the Upanishads, it placed the first importance on know-ledge; but it was no longer a knowledge of God, it was a clear perception of the real nature, as they supposed it to be, of men and things. And it added to the necessity of knowledge, the necessity of purity, of courtesy, of uprightness, of peace and of a universal love far reaching, grown great and beyond measure." (Rhys Davids, *Hibbert Lectures*, p. 29.)

The contemporary Greek philosophy takes an analogous direction. According to Heracleitus, the universe was made neither by Gods nor men; but, from all eternity, has been, and to all eternity, will be, immortal fire, glowing and fading in due measure. (Mullach, *Heracliti Fragmenta*, 27.) And the part assigned by his successors, the Stoics, to the knowledge and the volition of the ' wise man' made their Divinity (for logical thinkers) a subject for compliments, rather than a power to be reckoned with. In Hindu speculation the ' Arahat,' still more the 'Buddha,' becomes the superior of Brahma; the stoical 'wise man' is, at least, the equal of Zeus.

*

Berkeley affirms over and over again that no idea can be formed of a soul or spirit—" If any man shall doubt of the truth of what is here delivered, let him but reflect and try if he can form any idea of power or active being; and whether he hath ideas of two principal powers marked by the names of *will* and *understanding* distinct from each other, as well as from a third idea of substance or being in general, with a relative notion of its supporting or being the subject of the aforesaid power, which is signified by the name *soul* or *spirit*. This is what some hold: but, so far as I can see, the words *will, soul, spirit,* do not stand for different ideas or, in truth, for any idea at all, but for something which is very different from ideas, and which, being an agent, cannot be like unto or represented by any idea whatever [though it must be owned at the same time, that we have some notion of soul, spirit, and the operations of the mind, such as willing, loving, hating, inasmuch as we know or understand the meaning of these words"]. (*The Principles of Human Knowledge*, lxxvi. See also §§ lxxxix., cxxxv., cxlv.)

It is open to discussion, I think, whether it is possible to have 'some notion' of that of which we can form no 'idea.'

Berkeley attaches several predicates to the "perceiving active being mind, spirit, soul or myself" (Parts I. II.) It is said, for example, to be "indivisible, incorporeal, unextended, and incorruptible." The predicate indivisible, though negative in form, has highly positive consequences. For, if 'perceiving active being' is strictly indivisible, man's soul must be one with the Divine spirit : which is good Hindu or Stoical doctrine, but hardly orthodox Christian philosophy. If, on the other hand, the 'substance' of active perceiving 'being' is actually divided into the one Divine and innumerable human entities, how can the predicate 'indivisible' be rigorously applicable to it?

Taking the words cited, as they stand, they amount to the denial of the possibility of any knowledge of substance. 'Matter' having been resolved into mere affections of 'spirit,' 'spirit' melts away into an admittedly inconceivable and unknowable hypostasis of thought and power—consequently the existence of anything in the universe beyond a flow of phenomena is a purely hypothetical assumption. Indeed a pyrrhonist might raise the objection that if 'esse' is 'percipi' spirit itself can have no existence except as a perception, hypostatized into a 'self,' or as a perception of some other spirit. In the former case, objective reality vanishes; in the latter, there would seem to be the need of an infinite series of spirits each perceiving the others.

It is curious to observe how very closely the phraseology of Berkeley sometimes approaches that of the Stoics: thus (cxlviii.) " It seems to be a *general pretence of the unthinking* herd that *they cannot see God*......But, alas, we need only open our eyes to see the Sovereign Lord of all things with a more full and clear view, than we do any of our fellow-creatures......we do at all times and in all places perceive manifest tokens of the Divinity: everything we see, hear, feel, or any wise perceive by sense, being a sign or effect of the power of God" ......cxlix. " It is therefore plain, that *nothing can be more evident* to any one that is capable of the least reflection, *than the existence of God*, or a spirit who is intimately present to our minds, producing in them all that variety of ideas or sensations which continually affect us, on whom we have an absolute and entire dependence, in short, *in whom we live and move and have our being.*" cl. [But you will say hath Nature no share in the production of natural things, and must they be all ascribed to the immediate and sole operation of God?......if by *Nature* is meant some

being distinct from God, as well as from the laws of
nature and things perceived by sense, I must confess
that word is to me an empty sound, without any
intelligible meaning annexed to it.] Nature in this
acceptation is a vain *Chimœra* introduced by those
heathens, who had not just notions of the omni-
presence and infinite perfection of God."

Compare Seneca (*De Beneficiis*, iv. 7) :

* "Natura, inquit, hæc mihi præstat. Non intelligis
te, quum hoc dicis, mutare Nomen Deo? Quid enim
est aliud Natura quam Deus, et divina ratio, toti
mundo et partibus ejus inserta? Quoties voles tibi
licet aliter hunc auctorem rerum nostrarum com-
pellare, et Jovem illum optimum et maximum rite
dices, et tonantem, et statorem : qui non, ut historici
tradiderunt, ex eo quod post votum susceptum acies
Romanorum fugientum stetit, sed quod stant beneficio
ejus omnia, stator, stabilitorque est : hunc eundem et
fatum si dixeris, non mentieris, nam quum fatum
nihil aliud est, quam series implexa causarum, ille est
prima omnium causa, ea qua cæteræ pendent." It
would appear, therefore, that the good Bishop is
somewhat hard upon the 'heathen,' of whose words
his own might be a paraphrase.

There is yet another direction in which Berkeley's
philosophy, I will not say agrees with Gautama's, but
at any rate helps to make a fundamental dogma of
Buddhism intelligible.

" I find I can excite ideas in my mind at pleasure,
and vary and shift the scene as often as I think fit.
It is no more than willing, and straightway this or
that idea arises in my fancy : and by the same power,

it is obliterated, and makes way for another. This making and unmaking of ideas doth very properly denominate the mind active. This much is certain and grounded on experience. . . ." (*Principles*, xxviii.)

A good many of us, I fancy, have reason to think that experience tells them very much the contrary; and are painfully familiar with the obsession of the mind by ideas which cannot be obliterated by any effort of the will and steadily refuse to make way for others. But what I desire to point out is that if Gautama was equally confident that he could 'make and unmake' ideas—then, since he had resolved self into a group of ideal phantoms—the possibility of abolishing self by volition naturally followed.

Note 9 (p. 68).

According to Buddhism, the relation of one life to the next is merely that borne by the flame of one lamp to the flame of another lamp which is set alight by it. To the 'Arahat' or adept "no outward form, no compound thing, no creature, no creator, no existence of any kind, must appear to be other than a temporary collocation of its component parts, fated inevitably to be dissolved."—(Rhys Davids, *Hibbert Lectures*, p. 211.)

The self is nothing but a group of phenomena held together by the desire of life ; when that desire shall have ceased, "the Karma of that particular chain of lives will cease to influence any longer any distinct individual, and there will be no more birth; for birth, decay, and death, grief, lamentation, and despair will have come, so far as regards that chain of lives, for ever to an end."

The state of mind of the Arahat in which the desire of life has ceased is Nirvana. Dr. Oldenberg has very acutely and patiently considered the various interpretations which have been attached to ' Nirvana ' in the work to which I have referred (pp. 285 *et seq.*). The result of his and other discussions of the question may I think be briefly stated thus:

1. Logical deduction from the predicates attached to the term ' Nirvana ' strips it of all reality, conceivability, or perceivability, whether by Gods or men. For all practical purposes, therefore, it comes to exactly the same thing as annihilation.

2. But it is not annihilation in the ordinary sense, inasmuch as it could take place in the living Arahat or Buddha.

3. And, since, for the faithful Buddhist, that which was abolished in the Arahat was the possibility of further pain, sorrow, or sin; and that which was attained was perfect peace; his mind directed itself exclusively to this joyful consummation, and personified the negation of all conceivable existence and of all pain into a positive bliss. This was all the more easy, as Gautama refused to give any dogmatic definition of Nirvana. There is something analogous in the way in which people commonly talk of the ' happy release ' of a man who has been long suffering from mortal disease. According to their own views, it must always be extremely doubtful whether the man will be any happier after the ' release ' than

before. But they do not choose to look at the matter in this light.

The popular notion that, with practical, if not metaphysical, annihilation in view, Buddhism must needs be a sad and gloomy faith seems to be inconsistent with fact; on the contrary, the prospect of Nirvana fills the true believer, not merely with cheerfulness, but with an ecstatic desire to reach it.

Note 10 (p. 68).

The influence of the picture of the personal qualities of Gautama, afforded by the legendary anecdotes which rapidly grew into a biography of the Buddha; and by the birth stories, which coalesced with the current folk-lore, and were intelligible to all the world, doubtless played a great part. Further, although Gautama appears not to have meddled with the caste system, he refused to recognize any distinction, save that of perfection in the way of salvation, among his followers; and by such teaching, no less than by the inculcation of love and benevolence to all sentient beings, he practically levelled every social, political, and racial barrier. A third important condition was the organization of the Buddhists into monastic communities for the stricter professors, while the laity were permitted a wide indulgence in practice and were allowed to hope for accommodation in some of the temporary abodes of bliss. With a few hundred thousand years of immediate paradise in sight, the average man could be content to shut his eyes to what might follow.

Note 11 (p. 69).

In ancient times it was the fashion, even among the Greeks themselves, to derive all Greek wisdom from Eastern sources; not long ago it was as generally denied that Greek philosophy had any connection with Oriental speculation; it seems probable, however, that the truth lies between these extremes.

The Ionian intellectual movement does not stand alone. It is only one of several sporadic indications of the working of some powerful mental ferment over the whole of the area comprised between the Ægean and Northern Hindostan during the eighth, seventh, and sixth centuries before our era. In these three hundred years, prophetism attained its apogee among the Semites of Palestine; Zoroasterism grew and became the creed of a conquering race, the Iranic Aryans; Buddhism rose and spread with marvellous rapidity among the Aryans of Hindostan; while scientific naturalism took its rise among the Aryans of Ionia. It would be difficult to find another three centuries which have given birth to four events of equal importance. All the principal existing religions of mankind have grown out of the first three : while the fourth is the little spring, now swollen into the great stream of positive science. So far as physical possibilities go, the prophet Jeremiah and the oldest Ionian philosopher might have met and conversed. If they had done so, they would probably have disagreed a good deal; and it is interesting to reflect that their discussions might have embraced questions which, at the present day, are still hotly controverted.

The old Ionian philosophy, then, seems to be only one of many results of a stirring of the moral and intellectual life of the Aryan and the Semitic populations of Western Asia. The conditions of this general awakening were doubtless manifold; but there is one which modern research has brought into great prominence. This is the existence of extremely ancient and highly advanced societies in the valleys of the Euphrates and of the Nile.

It is now known that, more than a thousand — perhaps more than two thousand—years before the sixth century B.C., civilization had attained a relatively high pitch among the Babylonians and the Egyptians. Not only had painting, sculpture, architecture, and the industrial arts reached a remarkable development; but in Chaldæa, at any rate, a vast amount of knowledge had been accumulated and methodized, in the departments of grammar, mathematics, astronomy, and natural history. Where such traces of the scientific spirit are visible, naturalistic speculation is rarely far off, though, so far as I know, no remains of an Accadian, or Egyptian, philosophy, properly so called, have yet been recovered.

Geographically, Chaldæa occupied a central position among the oldest seats of civilization. Commerce, largely aided by the intervention of those colossal pedlars, the Phœnicians, had brought Chaldæa into connection with all of them, for a thousand years before the epoch at present under consideration.

And in the ninth, eighth, and seventh centuries, the Assyrian, the depositary of Chaldæan civilization, as the Macedonian and the Roman, at a later date, were the depositaries of Greek culture, had added irresistible force to the other agencies for the wide distribution of Chaldæan literature, art, and science.

I confess that I find it difficult to imagine that the Greek immigrants—who stood in somewhat the same relation to the Babylonians and the Egyptians as the later Germanic barbarians to the Romans of the Empire—should not have been immensely influenced by the new life with which they became acquainted. But there is abundant direct evidence of the magnitude of this influence in certain spheres. I suppose it is not doubted that the Greek went to school with the Oriental for his primary instruction in reading, writing, and arithmetic; and that Semitic theology supplied him with some of his mythological lore. Nor does there now seem to be any question about the large indebtedness of Greek art to that of Chaldæa and that of Egypt.

But the manner of that indebtedness is very instructive. The obligation is clear, but its limits are no less definite. Nothing better exemplifies the indomitable originality of the Greeks than the relations of their art to that of the Orientals. Far from being subdued into mere imitators by the technical excellence of their teachers, they lost no time in bettering the instruction they received, using their models as mere stepping stones on the way to those unsurpassed and unsurpassable achievements which are all their own. The shibboleth of Art is the human figure. The ancient Chaldæans and Egyptians, like the modern Japanese, did wonders in the representation of birds and quadrupeds; they even attained to something more than respectability in human portraiture. But their utmost efforts never brought them within range of the best Greek embodiments of the grace of womanhood, or of the severer beauty of manhood.

It is worth while to consider the probable effect upon the acute and critical Greek mind of the conflict of ideas, social, political, and theological, which arose out of the conditions of life in the Asiatic colonies. The Ionian polities had passed through the whole gamut of social and political changes, from patriarchal and occasionally oppressive kingship to rowdy and still more burdensome mobship—no doubt with infinitely eloquent and copious argumentation, on both sides, at every stage of their progress towards that arbitrament of force which settles most political questions. The marvellous speculative faculty, latent in the Ionian, had come in contact with Mesopotamian, Egyptian, Phœnician theologies and cosmogonies; with the illuminati of Orphism and the fanatics and dreamers of the Mysteries; possibly with Buddhism and Zoroasterism; possibly even with Judaism. And it has been observed that the mutual contradictions of antagonistic supernaturalisms are apt to play a large part among the generative agencies of naturalism.

Thus, various external influences may have contributed to the rise of philosophy among the Ionian Greeks of the sixth century. But the assimilative capacity of the Greek mind—its power of Hellenizing whatever it touched—has here worked so effectually, that, so far as I can learn, no indubitable traces of such extraneous contributions are now allowed to exist by the most authoritative historians of Philosophy. Nevertheless, I think it must be admitted that the coincidences between the Heracleito-stoical doctrines and those of the older Hindu philosophy are extremely remarkable. In both, the cosmos pursues an eternal succession of cyclical changes. The great year, answering to the Kalpa, covers an entire cycle from the origin of the universe as a fluid to its dissolution in fire—"Humor initium, ignis exitus mundi," as Seneca has it. In both systems, there is immanent in the cosmos a source of energy, Brahma, or the Logos, which works according to fixed laws. The individual soul is an efflux of this world-spirit, and returns to it. Perfection is attainable only by individual effort, through ascetic discipline, and is rather a state of painlessness than of happiness; if indeed it can be said to be a state of anything, save the negation of perturbing emotion. The hatchment motto "In Cœlo Quies" would serve both Hindu and Stoic; and absolute quiet is not easily distinguishable from annihilation.

Zoroasterism, which, geographically, occupies a position intermediate between Hellenism and Hinduism, agrees with the latter in recognizing the essential evil of the cosmos; but differs from both in its intensely anthropomorphic personification of the two antagonistic principles, to the one of which it ascribes all the good; and, to the other, all the evil.

In fact, it assumes the existence of two worlds, one good and one bad ; the latter created by the evil power for the purpose of damaging the former. The existing cosmos is a mere mixture of the two, and the ' last judgment' is a root-and-branch extirpation of the work of Ahriman.

Note 12 (p. 69).

There is no snare in which the feet of a modern student of ancient lore are more easily entangled, than that which is spread by the similarity of the language of antiquity to modern modes of expression. I do not presume to interpret the obscurest of Greek philosophers ; all I wish is to point out, that his words, in the sense accepted by competent interpreters, fit modern ideas singularly well.

So far as the general theory of evolution goes there is no difficulty. The aphorism about the river ; the figure of the child playing on the shore ; the kingship and fatherhood of strife, seem decisive. The ὁδὸς ἄνω κάτω μίη expresses, with singular aptness, the cyclical aspect of the one process of organic evolution in individual plants and animals : yet it may be a question whether the Heracleitean strife included any distinct conception of the struggle for existence. Again, it is tempting to compare the part played by the Heracleitean 'fire' with that ascribed by the moderns to heat, or rather to that cause of motion of which heat is one expression ; and a little ingenuity might find a foreshadowing of the doctrine of the conservation of energy, in the saying that all the

*

things are changed into fire and fire into all things,
as gold into goods and goods into gold.

Note 13 (p. 71).

Popes lines in the *Essay on Man* (Ep. i. 267–8),

> "All are but parts of one stupendous whole,
> Whose body Nature is, and God the soul,"

simply paraphrase Seneca's "quem in hoc mundo
locum deus obtinet, hunc in homine animus : quod
est illic materia, id nobis corpus est."—(Ep. lxv. 24) ;
which again is a Latin version of the old Stoical
doctrine, εἰς ἅπαν τοῦ κόσμου μέρος διήκει ὁ νοῦς,
καθάπερ ἀφ' ἡμῶν ἡ ψυχή.

So far as the testimony for the universality of what
ordinary people call 'evil' goes, there is nothing
better than the writings of the Stoics themselves.
They might serve as a storehouse for the epigrams of
the ultra-pessimists. Heracleitus (*circa* 500 B.C.)
says just as hard things about ordinary humanity
as his disciples centuries later ; and there really
seems no need to seek for the causes of this dark
view of life in the circumstances of the time of
Alexander's successors or of the early Emperors of
Rome. To the man with an ethical ideal, the world,
including himself, will always seem full of evil.

Note 14 (p. 73).

I use the well-known phrase, but decline respon-
sibility for the libel upon Epicurus, whose doctrines
were far less compatible with existence in a stye

than those of the Cynics. If it were steadily borne in mind that the conception of the ' flesh ' as the source of evil, and the great saying 'Initium est salutis notitia peccati,' are the property of Epicurus, fewer illusions about Epicureanism would pass muster for accepted truth.

*

Note 15 (p. 75).

The Stoics said that man was a ζῷον λογικὸν πολιτικὸν φιλάλληλον, or a rational, a political, and an altruistic or philanthropic animal. In their view, his higher nature tended to develop in these three directions, as a plant tends to grow up into its typical form. Since, without the introduction of any consideration of pleasure or pain, whatever thwarted the realization of its type by the plant might be said to be bad, and whatever helped it good ; so virtue, in the Stoical sense, as the conduct which tended to the attainment of the rational, political, and philanthropic ideal, was good in itself, and irrespectively of its emotional concomitants.

Man is an " animal sociale communi bono genitum."
The safety of society depends upon practical recognition of the fact. "Salva autem esse societas nisi custodia et amore partium non possit," says Seneca. (*De. Ira*, ii. 31.)

†

‡

Note 16 (p. 75).

The importance of the physical doctrine of the Stoics lies in its clear recognition of the universality of the law of causation, with its corollary, the order of nature : the exact form of that order is an altogether secondary consideration.

Many ingenious persons now appear to consider that the incompatibility of pantheism, of materialism, and of any doubt about the immortality of the soul, with religion and morality, is to be held as an axiomatic truth.   I confess that I have a certain difficulty in accepting this dogma.   For the Stoics were notoriously materialists and pantheists of the most extreme character ; and while no strict Stoic believed in the eternal duration of the individual soul, some even denied its persistence after death.   Yet it is equally certain that of all gentile philosophies, Stoicism exhibits the highest ethical development, is animated by the most religious spirit, and has exerted the profoundest influence upon the moral and religious development not merely of the best men among the Romans, but among the moderns down to our own day.

Seneca was claimed as a Christian and placed among the saints by the fathers of the early Christian Church ; and the genuineness of a correspondence between him and the apostle Paul has been hotly maintained in our own time, by orthodox writers. That the letters, as we possess them, are worthless forgeries is obvious ; and writers as wide apart as Baur and Lightfoot agree that the whole story is devoid of foundation.

The dissertation of the late Bishop of Durham (*Epistle to the Philippians*) is particularly worthy of study, apart from this question, on account of the

*

evidence which it supplies of the numerous similarities of thought between Seneca and the writer of the Pauline epistles. When it is remembered that the writer of the Acts puts a quotation from Aratus, or Cleanthes, into the mouth of the apostle; and that Tarsus was a great seat of philosophical and especially stoical learning (Chrysippus himself was a native of the adjacent town of Sôli), there is no difficulty in understanding the origin of these resemblances. See, on this subject, Sir Alexander Grant's dissertation in his edition of *The Ethics of Aristotle* (where there is an interesting reference to the stoical character of Bishop Butler's ethics), the concluding pages of Dr. Weygoldt's instructive little work *Die Philosophie der Stoa*, and Aubertin's *Sénèque et Saint Paul*.

It is surprising that à writer of Dr. Lightfoot's stamp should speak of Stoicism as a philosophy of 'despair.' Surely, rather, it was a philosophy of men who, having cast off all illusions, and the childishness of despair among them, were minded to endure in patience whatever conditions the cosmic process might create, so long as those conditions were compatible with the progress towards virtue, which alone, for them, conferred a worthy object on existence. There is no note of despair in the stoical declaration that the perfected 'wise man' is the equal of Zeus in everything but the duration of his existence. And, in my judgment, there is as little pride about it, often as it serves for the text of discourses on stoical arrogance. Grant the stoical postulate that there is no good except virtue; grant that the perfected wise man is altogether virtuous, in consequence of being guided in all things by the reason, which is an effluence of Zeus, and there seems no escape from the stoical conclusion.

Note 17 (p. 76).

Our "Apathy" carries such a different set of connotations from its Greek original that I have ventured on using the latter as a technical term.

Note 18 (p. 77).

Many of the stoical philosophers recommended their disciples to take an active share in public affairs; and in the Roman world, for several centuries, the best public men were strongly inclined to Stoicism. Nevertheless, the logical tendency of Stoicism seems to me to be fulfilled only in such men as Diogenes and Epictetus.

Note 19 (p. 80).

"Criticisms on the Origin of Species," 1864. *Collected Essays,* vol. ii. p. 91. [1894.]

Note 20 (p. 81).

Of course, strictly speaking, social life, and the ethical process in virtue of which it advances towards perfection, are part and parcel of the general process of evolution, just as the gregarious habit of in-

numerable plants and animals, which has been of
immense advantage to them, is so. A hive of bees
is an organic polity, a society in which the part
played by each member is determined by organic
necessities. Queens, workers, and drones are, so to
speak, castes, divided from one another by marked
physical barriers. Among birds and mammals,
societies are formed, of which the bond in many cases
seems to be purely psychological ; that is to say, it
appears to depend upon the liking of the individuals
for one another's company. The tendency of
individuals to over self-assertion is kept down by
fighting. Even in these rudimentary forms of society,
love and fear come into play, and enforce a greater
or less renunciation of self-will. To this extent the
general cosmic process begins to be checked by a rudimentary ethical process, which is, strictly speaking,
part of the former, just as the 'governor' in a steam-
engine is part of the mechanism of the engine.

Note 21 (p. 82).

See "Government : Anarchy or Regimentation,"
Collected Essays, vol. i. pp. 413—418. It is this
form of political philosophy to which I conceive
the epithet of 'reasoned savagery' to be strictly
applicable. [1894.]

Note 22 (p. 83).

"L'homme n'est qu'un roseau, le plus faible de
la nature, mais c'est un roseau pensant. Il ne faut

pas que l'univers entier s'arme pour l'écraser. Une vapeur, une goutte d'eau, suffit pour le tuer. Mais quand l'univers l'écraserait, l'homme serait encore plus noble que ce qui le tue, parce qu'il sait qu'il meurt ; et l'avantage que l'univers a sur lui, l'univers n'en sait rien."—*Pensées de Pascal.*

Note 23 (p. 85).

The use of the word " Nature " here may be criticised. Yet the manifestation of the natural tendencies of men is so profoundly modified by training that it is hardly too strong. Consider the suppression of the sexual instinct between near relations.

Note 24 (p. 86).

A great proportion of poetry is addressed by the young to the young ; only the great masters of the art are capable of divining, or think it worth while to enter into, the feelings of retrospective age. The two great poets whom we have so lately lost, Tennyson and Browning, have done this, each in his own inimitable way ; the one in the *Ulysses*, from which I have borrowed ; the other in that wonderful fragment ' Childe Roland to the dark Tower came.'

Editors' Notes

[p. 11] * *Collected Essays*, vol. 9, 195–244.

[p. 40] * "The Struggle for Existence in Human Society," in *Collected Essays*, vol. 9, 195–244.

[p. 46] * "I like to cross over into the enemy camp—not as a deserter, but as a spy."

[p. 89] * "For it is not possible to step twice into the same river."

      † "Our bodies are rushed along like flowing rivers. Whatever you see, flows with time. Nothing of the things we see is fixed. I myself, as I talk about this change, am changed. This is what Heraclitus says: 'We go down into the same river twice, and yet we don't go down [into the same river].' For the river keeps the same name, but the water has flowed on. Of course, this is more obvious in rivers than in man, but still we are carried along in no less rapid course."

      ‡ "Many of our blessings hurt us; memory recalls the tortures of fear, and foresight anticipates them. No one is wretched in present circumstances."

[p. 91] * Thomas William Rhys Davids, *Lectures on the Origin and Growth of Religion as Illustrated by Some Points in the History of Indian Buddhism*, Hibbert Lectures (London: Williams and Norgate, 1881), 83.

      † Thomas William Rhys Davids, *Buddhism: Being a Sketch of the Life and Teachings of Gautama, the Buddha* (London: Society for Promoting Christian Knowledge, 1890).

      ‡ Hermann Oldenberg. *Buddha: Sein leben, sine lehre, seine gemeinde*, 2d ed. (Berlin: W. Hertz, 1890).

[p. 96] * "If anything except living honorably is good, we will be pursued by a greed for life, and by greed for the furnishings of life—which is intolerable, infinite, unstable."

[p. 97] * Friedrich Wilhelm August Mullach, ed., "Heracliti Fragmenta," in *Fragmenta philosophorum Graecorum* . . . , (Paris: Firmin Didot, 1860–81), 1:27.

[p. 99]   *   " 'to be' is 'to perceive' "

[p.100]   *   "Nature, you say, presents me with these things. But don't you
see that when you say that, you are just giving another name to
God? What else is Nature but God and divine reason that is
everywhere in the whole universe and all its parts? You may
address this being who is the author of our world as often as
you like by different names. You can call him Jove the Best and
the Greatest, as well as Thunderer and the Stayer—not because, as historians tell us, the Roman front [battle line] stayed
its flight in answer to prayer, but because everything is upheld
by his benefits—he is their supporter and stabilizer. If you call
him Fate, you wouldn't be wrong, since fate is nothing else but
a connected chain of causes, and he is the first cause on which
the others depend."

[p.108]   *   "Water is the beginning, fire the end of the world."

           †   In Heaven, Quiet

[p.109]   *   "The road up and the road down are one."

[p.110]   *   "The place that God holds in the world is like the soul in man.
What is material in that [in the world] is body in us."

           †   "Mind winds through every part of the cosmos just as the soul
in us."

[p.111]   *   "The beginning of salvation is the awareness of sin."

           †   "social animal, born to the common good"

           ‡   "For a society is not able to be healthy, except through the care
and love of its parts."

[p.112]   *   Joseph Barber Lightfoot, *St. Paul's Epistle to the Philippians: a
Revised [Greek] Text* (London: Macmillan, 1868).

[p.113]   *   Alexander Grant, *The Ethics of Aristotle Illustrated with Essays and
Notes*, 3 vol. (London: J. W. Parker, 1857–58); Georg P. Wey-
goldt, *Die Philosophie der Stoa, nach ihren Wesen und ihren Schick-
salen* (Leipzig: O Schulze, 1883); Charles Aubertin, *Sénèque et
Saint Paul: Étude sur les Rapports Supposés entre le Philosophe et
l'Apôtre* (Paris: n.p., 1869).

[p.115]   *   "Man is but a reed, the most feeble thing in nature; but he is a
thinking reed. The entire universe need not arm itself to crush

176

him. A vapor, a drop of water suffices to kill him. But if the universe were to crush him, man would still be more noble than that which killed him, because he knows that he dies and the advantage which the universe has over him; the universe knows nothing of this" [*Pascal's Pensees* (New York: E. P. Dutton, 1958)].

A Sociobiological Expansion of
Evolution and Ethics

♦

GEORGE C. WILLIAMS

H UXLEY WROTE almost a century ago, and since his time there has been a great advance in our understanding of the process of organic evolution and a great increase in our fund of knowledge about its products. My purpose here is to review those aspects of current theoretical understanding and biological knowledge that bear most forcefully on the main theme of Huxley's essay, his moral evaluation of nature. The theoretical specifics will no doubt prove temporary. Graphical and symbolic models of sexual selection, for example, will be quite different in a few decades from any in use today. I expect greater permanence for the empirical summaries I present; estimates of average rates of infanticide in ground squirrel colonies, for example, are likely to need but minor adjustments.

In this commentary I do not hesitate to draw philosophical conclusions from current biological ideas and data, but must warn that my perspective and training are those of a biologist, not an ethical philosopher. I must also stress that the selection of ideas and illustrations is my own, and that another biologist would surely choose a different range of topics and reach at least somewhat different conclusions. I can also claim that there is nothing idiosyncratic or unorthodox about the biology represented here.

Huxley summarizes his moral evaluation of nature succinctly:

> Thus, brought before the tribunal of ethics, the cosmos might well seem to stand condemned. The conscience of man revolted against the moral indifference of nature, and the microscopic atom should have found the illimitable macrocosm guilty. (p. 59)

This is an indecisive and disappointing statement, by my reading of the biological macrocosm. I am inclined to substitute *gross immorality* for *moral indifference* and to avoid the subjunctive in the final clause. I attribute the weakness of Huxley's version mainly to the weakness of his grasp of natural selection, even in its nineteenth-century form. His grasp was inferior to that of Darwin and a few other nineteenth-century biologists like Wallace, Weismann, and D. S. Jordan, as has often been noted (Leonard Huxley 1900, 12, 372; Mayr and Provine, 1980, 133). No one of Huxley's generation could have imagined the current concept of natural selection, which can honestly be described as a process for maximizing short-sighted selfishness. I would concede that moral indifference might aptly characterize the physical universe. For the biological world a stronger term is needed.

My claim that natural selection and its resulting biological effects are immoral may be challenged on the same basis as Dawkins (1976) was challenged on his use of *selfish* to describe the gene. If morality and selfishness must connote an awareness of evil or any degree of personal accountability, such terms are indeed inappropriate for any entity with a markedly less-than-human level of understanding. I find this terminological constraint inconvenient, however, and I think it contrary to much normal usage. My use of *immoral* is like that of a pacifist who maintains that war is immoral. Such a claim may imply that Mars plants evil thoughts in the minds of war lords, and that they and he rejoice in the clatter of bayonets, but this would not normally be understood. Likewise, the pacifist's realization that wars start for complex reasons, and seldom merely from anyone's consciously evil decision, need not make him prefer Huxley's *moral indifference* to my *immorality*. A pacifist regards war as immoral because it frequently and systematically produces results that he finds morally unacceptable.

Surely there is a morally important difference between being struck by lightning and being struck by a rattlesnake. No one could fail to see that the rattlesnake has what are clearly weapons, precisely designed and used so as to produce a victim. The distinction between pain and death caused by aggressive use of weapons and pain and death caused by accident is surely worth maintaining, but it is obscured when such terms as *moral indifference* are applied to both. It illustrates a decisive contrast between biological nature and physical nature, one that Huxley did not fully appreciate. Of course I concede both the greater evil of murder by a human being over predation or lethal self-defense by a rattlesnake and the absurdity of holding the reptile accountable for its actions. There is a major difference between a human mind and a snake mind, but this is already widely appreciated, and I think there is a greater need to emphasize the other sort of difference, between the organization and behavior of a rattlesnake and those of lightning. I hope that my discussion will show, at least to those who would concede that war is or may be immoral, that organic evolution is worse than traditional forms of warfare, and worse than Huxley imagined.

RECENT ADVANCES IN THE UNDERSTANDING OF NATURAL SELECTION

The term *natural selection* is used with sometimes a more and sometimes a less inclusive meaning, usually apparent from context. Sometimes it refers to biases affecting success and failure for organisms in nature: an early bird gets the worm, a later one does not; one worm gets eaten; a different one escapes; one kind of organism establishes itself in a new habitat, another kind dies out. In other discussions the term focuses on the keeping of records on such events: this means changes in gene frequency. The eating of worms by early birds and deprivation of late risers will affect evolution only if these processes affect gene frequencies in the bird population. Regularities in success and failure and their recording by gene frequency changes are both essential to evolution by natural selection. One should be clear as to whether one or the other or both are being discussed, but it would be nonsense to argue their relative importance.

Confusion on this point obfuscates some recent discussion of levels of selection. It is legitimate to ask whether selection in a species of bird acts mainly through successes and failures of different individuals in each population (individual selection), or by successes and failures of local populations in maintaining themselves and spreading to other localities (group selection). To argue about whether selection is more important at the level of the individual or the gene (e.g., Gould 1982) makes as much sense as claiming that either Boswell or Johnson was more important for Boswell's *Life of Johnson*. Recognition of the distinction between events taking place and the record being kept had to await a thorough appreciation of the relationship between gene and character, and between genotype and phenotype. Neither the distinction nor its moral implications could have been comprehended in Huxley's time.

The distinction is essential to any understanding of the role recognized by modern biology for individual organisms in Huxley's cosmic process. The gene pool of a population is a record of reproductive success and failure in that population, and at conception an organism gets a sample of this record. The sample is its instructions for producing the machinery by which it adapts to its environment. All other useful information, such as that learned and stored in the brain, depends on the initial genetic information. New information can be exploited because organs for the gathering, storage, and use of information are specified in the genes. Such organs are no less biological than those for the gathering, storage, and use of food. Both kinds of organs are there because they have been useful in previous generations for transmitting genes to later generations. The maximal achievement of this genetic success is the role of the individual in evolution.

This is as good a statement as I can make in a few sentences on the evolutionary significance of the individual, according to current consensus among evolutionary theorists. Of course it is oversimplified. For instance, it incorrectly implies that selection of genes depends entirely on their effects on developed phenotypes. Geneticists have recently pointed out subtle ways by which genes can increase their rates of transmission without affecting the survival or reproduction of the individuals in which they occur. Such insights may lead to answers to some fundamental questions, such as why reproduction is so often sexual and inheritance Mendelian. I prefer to ignore such matters here, because the better known phenotype-mediated selection has clearer relevance to ethics.

The fact that inheritance is Mendelian justifies the reductionist view that Darwinian fitness is ultimately a property of that which obeys Mendel's law of independent assortment. Such an entity is a gene by the definition implied in most discussions of population genetics. Two DNA molecules that make two different products would be a single gene in this sense if they are so closely linked that they fail the criterion of independent transmission even after many generations. To a physiological or developmental geneticist, the two molecules might be two different genes. In most formulations of natural selection, a Y chromosome would be a single maleness gene in the human and many animal populations. Any consistent bias in rates of transmission of distinguishable Y chromosomes would measure the fitness of these competing Ys. Fitness can be a property of whole individuals (or groups thereof) only secondarily and with qualifications.

This reductionistic view has been challenged by both philosophers and biologists (e.g., Wimsatt 1980; Gould 1980, 1982) and defended by others (Ruse 1986, chapter 1; Dawkins 1982). Few facts or biological concepts are in dispute. Wimsatt and Gould recognize that every human organism now alive will die in either the twentieth or twenty-first century, and that no one can transmit either a genotype or phenotype to the next generation. All anyone can do is to furnish each gamete with a randomly selected set of genes that can form a new and unique genotype by combination with a set from another individual. The only biological continuity between people today and those of the twenty-second century will be their genes. We have much in common with people who may be living then, just as we have with those of the eighteenth century, and the recurrence of common features, biological or cultural, is of highest importance. It is no less important that individuals and their visible features recur rather than persist. The repeated recurrence of mortal individuals with their temporary genotypes and phenotypes is made possible by the repeated transmission of immortal genes.

These facts of life are a set of conceptual constraints that must form part of the discipline of evolutionary biologists in their study of organisms or populations or more inclusive entities. Such a gene-centered view of adaptation in no way diminishes the importance of the higher-level studies. If the statement that the gene must be the basic unit of fitness and selection leads anyone to believe that an understanding of the formalities of population genetics is equivalent to an understanding of evolution, fears that this view is misleading would be justified. I hope that my distinction between regularities in events in nature and the genetic mode of their recording will help to avoid the dangers of reductionism while exploiting its conceptual convenience.

Population genetics as pioneered by Fisher, Haldane, and Wright around 1930 provides the basis for the current view of the record-keeping aspect of natural selection. A corollary of this advance, especially important for social behavior and its moral implications, was first analyzed in detail by Hamilton (1964). He pointed out that the survival and reproduction of a relative are partly equivalent, in evolutionary effect, to one's own survival and reproduction. He proposed an expansion of Darwin's concept of fitness to include any ability to get one's own genes represented in future generations, whether these genes are present in one's own cells or present, for genealogical reasons, in relatives. He termed this broader concept *inclusive fitness*. His key formulation was that assistance to a relative will be favored if the benefit to the relative, times the coefficient of relationship, exceeds the cost to the donor. Costs and benefits are always measured in the currency of reproductive success.

This insight is now an essential element in our understanding of social relationships in animals and has recently found applications with cellular interactions in microorganisms (Janzen 1977) and plants (Queller 1983; Willson and Burley 1983). Consider the problem of maximizing the fitness of individual 5 in the diagram below. She is fit according to her ability to get her genes passed on to future generations. This special set of genes is found not only in herself but to varying degrees in five of the other individuals, as indicated by the shading.

Solving this problem requires answering such questions as: what is the probability that an arbitrary chosen gene in the focal individual (5) is also present in her father (1)? In sexual reproduction half of one's genes come from one's father. So the probability is .5, and this is the coefficient of relationship of individual 1 to his daughter 5. The gene would be favored if it tended in any way to cause 5 to behave towards her father as if his genes were half as important as hers. Her inclusive fitness would be enhanced, up to a point, by a willingness to jeopardize her own well-being to increase his.

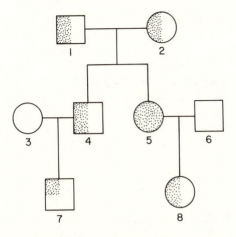

Hypothetical pedigree diagram, applicable to any outcrossed diploid organism with separate sexes. Males are represented by squares, females by circles, matings by horizontal lines connecting males and females, offspring by vertical connections. Stippling shows the proportion of the genes in each individual that are identical by descent with those in individual 5. Her fitness is measured by her ability to propagate these genes. Unstippled regions represent random samples of genes on which 5 can have only random effects.

The extent of the self-sacrifice favored would be determined by the .5 coefficient and the cost-benefit relationship. All costs and benefits would be measured as effects on the likelihood of successful reproduction. If we assume exactly equal prospects for future reproduction by 5 and 1 and absolute certainty that 1 is 5's father, natural selection would favor her risking her life up to a .5 probability of loss, if she can surely save him by taking the risk, and if he will surely die if she does not. When it is nephew 7 who is threatened under otherwise similar conditions, .25 would be the maximally tolerable risk. Selection based on this conditional advantage of aid to relatives was termed *kin selection* by Maynard Smith (1964).

Kin selection normally deals with less than life-and-death matters. A more likely problem is whether to give a food item to a nephew (*7*) or a daughter (*8*). The coefficient of relationship to a daughter is twice what it is to a nephew, so the nephew's potential benefit would have to be twice that of a daughter to justify giving it to him. Selection always favors nepotism towards close relatives over distant ones. In the symbolism of the diagram, the evolutionary role of individual *5* is to devote her life to the proliferation of that represented by the stippling. That which assigns this role to human individuals as to any other organism is what Huxley called "a tenacious and powerful enemy" (p. 85).

Some Common Misunderstandings

I have tried to give as fair a view of current understanding of kin selection as can be presented in a few paragraphs. I also realize that specialists in any subject may use common terms in uncommon ways, and that understanding by nonspecialists depends on much mind reading and intuition. For instance, when I discussed how an individual ought to behave towards her father, I could not mean literally that. I meant how she could be expected to react to stimuli associated in any way with fatherhood. Neither a wild animal nor a hunter-gatherer knows anything of coefficients of relationship or of roles assigned by a tenacious and powerful enemy. Sociobiologists often use arguments like this:

Suppose two male rats are fighting over a female and the stronger disables the weaker. The victor can now kill his rival, but he has a gene for altruism that makes him allow the rival to slink away. The rival may then recover, reassert himself, and perhaps defeat the former victor. The gene for altruism can thus be a disadvantage for its bearer, but this need not mean that it would be selected against. Perhaps the population structure is such that rivals can have an appreciable coefficient of relationship, maybe .1 on average (a bit less than first cousins). Killing the rival would, on average, be equivalent to killing ten percent of oneself. If this loss is greater than that expected later from a defeated rival, the altruism gene would be favored and replace an alternative allele for killing a rival.

Several features of this argument need comment. It assumes the point made above, that coefficients of relationship can influence behavior only indirectly through associated stimuli. It is hypothesized that a rat population has an appreciable viscosity, so that two adult males in the same neighborhood are likely to have some recent genealogical connection. The mere fact of sexual rivalry would then be predictive of an appreciable degree of kinship.

All favoritism towards relatives depends on associated observations that make it possible to discriminate differing degrees of kinship. A male songbird observes that his mate lays eggs in his nest and they are replaced later by hungry nestlings. Kin selection will favor his reacting to each of the young as if it were genetically equivalent to a fraction of himself. The fraction would be a product of the average father-offspring relationship (a bit more than .5 if there is appreciable inbreeding), times the confidence of paternity (he inseminated his mate, but he cannot be sure that he is the only male that did), times the confidence of maternity (a neighboring female may have stealthily left one or more of the eggs). Such nest parasitism by other species, cowbirds, for instance, is well known, and it is more common within species than is generally realized (Andersson and Eriksson 1982). Maybe the average coefficient of relationship between males and young in their nests in recent evolutionary time in this population is .40 and that for females .45. A slightly lower level of kin-selected altruism would then be expected for males than for females.

The appearance of the young and their presence in a particular nest are important stimuli that trigger the reactions of the parents. Only as long as this information is reliably predictive of special genetic similarity will the onerous task of feeding such young be favored. Recent evidence suggests that some animals make use of many subtle kinds of evidence in their assessments of kinship, and make comparisons between supposed relatives or even with themselves (Blaustein 1983; Lacy and Sherman 1983; Massey 1977).

An aspect of the rat story that might be confusing is the calling forth of a gene for a certain kind of altruistic behavior. The statement "he has a gene for altruism" may actually have several levels of meaning, one merely an identification with current orthodoxy. Only recently has there been a general consensus among biologists on the mechanisms of evolution. When I was in college in the 1940s, many professional biologists subscribed to diverse forms of orthogenesis, Lamarckism, hopeful-monsterism, and other views now considered obsolete. With the word *gene* one subtly waves the neo-Darwinian banner.

The gene-for-altruism terminology may also imply more than is intended. No biologists believe that there is a gene locus with some ordained role in adjusting altruism among rats. They merely believe that some quantitative aspect of the behavior, such as the probability or intensity of attack on a beaten opponent, is subject to some genetic variation. This belief is more credible for a complex end product of development, such as social behavior, than for those closer to primary gene action, such as enzyme structure. An attack on a defeated rival requires a host of sensory devices, perhaps mainly vision and receptors in muscles and joints. It requires the complex motor machinery of many bones and muscles. It depends on many neural and hormonal mechanisms that govern motivation, information processing, and motor control. Even a minor mutational change in any of these structural, physiological, or psychological properties would be unlikely to have no effect at all on some complex behavior pattern. The neo-Darwinian shorthand *a gene for* is a general reference to any one of a presumably large number of loci at which current variation has at least a minute effect on the degree of development of the character being considered.

A biologist attributes behavioral effects to a hypothetical gene for the same reason that a chemist postulates combining properties for an atom. Such properties are an essential part of the microstructure of the chemist's account of a reaction, but not the whole story. If the logic makes sense for one element of the microstructure (atom or gene) it should make sense for others. An argument about a gene for altruism tacitly assumes an extension such as *ditto for all other altruism loci.*

In another sense the rat story goes too far. The conclusion that a kill-your-opponent gene is entirely replaced by a let-him-go gene is extreme and improbable, but may be included to cover all relevant possibilities. If a gene frequency can shift from low to complete fixation as a result of favorable selection over the whole range, it can also go from .05 to .10 or .70 to .80 with favorable selection over these limited ranges.

The actual evolution of the rat population might be something like this. If conditions have been nearly stable for a while, there should be a nearly optimum level of belligerency among competing males. Sometimes the winners kill the losers, and sometimes they let them escape. An environmental change then somehow curtails dispersal so that rivals tend to be more closely related than they used to be. Selection will now act, at many loci and through a variety of developmental mechanisms, to increase the likelihood of the winner letting the loser escape. Loci at which genetic differences have the greatest effect on this likelihood, and those at which allelic frequencies are most equitable, will be most strongly affected. Very little of the evolutionary change need depend on mutations that arise after the change in conditions. Preexisting variation would usually furnish adequate raw material. I would also assume that most of the genetic changes would be minor quantitative shifts, such as .05 to .10, rather than fixation of an initially rare and loss of the previously normal allele.

I will mention one more of the many possible misunderstandings. Even a complete kin-selection argument would not give a full account of the natural selection at work on a social trait. Kinship is merely one of many important factors. The individuals in the diagram must vary in many ways, not merely in kinship to the one I picked for special attention. That individual may have established mutually beneficial reciprocation with others, some perhaps unrelated. Her mate (6) would be an obvious example. Her brother's mate (8) is also unrelated, but even if this female has no other significance, she has value (to 1) if she is helping nephew 7. Individuals also vary in the amount of benefit they would realize from a given donation. A postmenopausal mother and adolescent daughter are equally related to oneself, but it obviously makes evolutionary sense to be more generous to the young than to the old relative. Questions such as why animals concern themselves so much more with their offspring than with equally related brothers and sisters are based on the assumption that kinship, to be important in evolution, has to be the only factor. A complete answer to such a question would have to consider the relative reliability, for adults of reproductive age, of evidence for parenthood and for other kinds of relationship; the reproductive potential, likely survival, and needs of recipients; needs and capabilities of other possible donors; and alternatives for the investment of resources.

RECENT EVOLUTIONARY THOUGHT AND THE MORALITY OF HUXLEY'S COSMIC PROCESS

It is only a quarter of a century since Hamilton (1964) provided the basis for understanding kinship as a factor in the evolution of social behavior, and less than that since Trivers's (1971) insights on conditions for the evolution of reciprocity. Prior to these developments there was no tradition of conceptual rigor in interpreting what animals of the same species were doing with each other. In the absence of explicit theory, the apparently benign and cooperative nature of some social interactions made it possible to paint a morally acceptable picture of nature. Romantically inspired naturalists could find homilies in the mutual grooming of monkeys, the feeding of one bird by another, the stationing of bull ungulates between predators and the more vulnerable females and young, and the brave self-sacrifice of a worker bee for colony and queen.

Such observations were seen as parables for moral guidance in a system that Campbell (1978) called *normative biologism*, perhaps a modern form of what Huxley recognized as the "immeasurable mischief" caused by "the moralizing of sentimentalists" (p. 74). The noble-savage tradition in romantic literature is a part of this larger complex. The first episode of the widely acclaimed television series "Roots," with its Arcadian view of African tribal society, shows that the tradition is currently alive and well. It persists despite abundant evidence of the personal cruelty, group bigotry, and environmental destructiveness of primitive human society (Day 1953; Guthrie 1971; Villa 1986; Chagnon 1988).

Romantics could deal with any selfishness and cruelty in nature by name-changing and selective attention. Territorial disputes could be seen, not as truculent striving to garner resources for oneself at one's neighbor's expense, but as a wise program of birth control through limitation of the number of breeding territories. Verbal camouflage is not merely practiced, but may be explicitly advocated. Objections have recently been made to accusing animals of practicing slavery, adultery, rape, or other sins (Estep and Bruce 1981; Gowaty 1982; Leacock 1980). The usual form of argument urges that some human practice, such as slavery, is largely determined by culture rather than heredity and differs in many descriptive details from anything practiced by ants. So one must not use the term slavery for any kind of ant behavior. These arguments are never used for such terms as *courtship, singing,* and other culture-laden and descriptively unique human activities. The essential requirement for the objections is for the term to mean something wicked, for it is the implication of wickedness in nature, rather than any aversion to anthropomorphism, that motivates the objections.

It is understandable if not logically defensible that even highly disciplined biologists would rather study what they find appealing than what they find less so. There is a much larger literature on feeding than on defecation. Biologists also spend more time studying defense against predators than against parasites, even though animals spend much more time scratching at small enemies than fleeing from large ones. Occasionally a decision to ignore unpleasant behavior is clearly stated. Blaffer Hrdy (1977, 3) found a superb example in a detailed study of antarctic penguins by Murray Lavick, who stated that "many of the colonies are plagued by little knots of 'Hooligans' who hang about their outskirts, and should a chick go astray it stands a good chance of losing its life at their hands. The crimes that they commit are such as to find no place in this book."

Times have changed. Biologists today realize that unpleasant behavior can be important, and they are increasingly willing to study it. They often conclude that unpleasantness may be normal and adaptive, and they are less inclined to excuse it as pathological or the result of unusual conditions. They are also reinterpreting various seemingly benign or cooperative sorts of behavior. Armed with insights from Hamilton and Trivers, they usually find that altruistic behavior is limited to special situations in which it can be explained by one or more of three possible factors, none of any use as a romantic's "exemplar for human conduct" (pp. 73–74).

The first and most prevalent is the nepotism explicitly discussed by Hamilton (1964). Cues available to a donor indicate an expected coefficient of relationship with a recipient, and the value of the aid given, times the coefficient or relationship, exceeds the cost. The service rendered is an investment by the donor in its own genes, as these are represented in the recipient. As the conditions differ, so does the behavior. Full sibs are nicer to each other than half sibs. Mother love seems to be greater for more nearly mature offspring, but less in younger than in older mothers (Anderson et al. 1980; Bierman and Robertson 1981). Anyone who feels that this sort of quantification destroys the beauty of family life is a modern equivalent of those who felt that Newtonian optics destroyed the beauty of the rainbow (Grunbaum 1952).

Cost-benefit relations, kinship effects, and information requirements are equally applicable to the bringing home of food for young, to their feeding by lactation, and to their feeding by exchanges across a placenta. They are applicable to placental exchanges between a plant and its seeds. Information indicative of kinship can be of varying reliability, from near certainty of maternity in viviparous animals to much lower levels of reliability for other sorts of relationship, such as sibhood. The production of misinformation on kinship as on other matters can be normal and adaptive, as indicated by the next factor.

Manipulation is the second reason why one individual may provide benefits for another. A mouse can provide nutritive benefits to a cat, and when handled by the cat's paws is literally manipulated, but the term normally has a metaphorical meaning. For instance, the threat of physical force may substitute for force itself, as when a subordinate monkey relinquishes a feeding site to a dominant one. The subordinate's attempt to maintain occupancy might bring worse harm from the dominant than the loss of food. More interesting kinds of manipulation result from deception, most obvious when practiced against another species. A snapper may swim eagerly to the jaws of an anglerfish that deceives it with a lure. Social-insect pheromones that suppress reproduction by workers are interpreted either as a kin-selected response by workers, or as manipulation by the queen, or perhaps a combination of both. Dawkins (1982) discusses this problem and gives a thoughtful and convincing argument for the general importance of manipulation as a factor in animal behavior. There should be no need to document the prevalence of manipulation by verbal messages in our own species. The importance of skill in such manipulation and in avoiding its adverse effects must have been important factors in the evolution of linguistic ability.

The term *manipulation* implies exploitation by a manipulator, but it may be that communication within a species is usually adaptive for both the sender and receiver of a message. A hen may use vocal signals to get her chicks to behave in ways that serve her genetic interests, but it will usually serve a chick's interest to obey. The same might be said for most verbal exchanges between human parents and their children, and manipulation may have even broader positive effects in human society. Anyone who makes an anonymous donation of money or blood or other resources as a result of some public appeal is biologically just as much a victim of manipulation as the snapper in the jaws of the anglerfish. In an otherwise enlightened treatment of ethics and sociobiology, Singer (1981) misses the importance of manipulation and its role in philanthropy and social activism. He maintains that the prevalence of anonymous blood donation shows an altruism in human nature beyond what would be expected from natural selection.

The third reason for one individual acting in the service of another is reciprocity. Whatever is given up by the donor costs it less than some repayment expected from the recipient. Neither the expectation nor the repayment need be conscious, or even behavioral. The secretions of an aphid may later repay an ant that protects it from a predator. In cooperative ventures the mutual benefits may be simultaneous. Perhaps two wolves in a cooperative attack on a large deer are more than twice as likely to succeed as one acting alone, and one deer can feed many wolves. A cooperative wolf may therefore eat while a loner goes hungry. We all routinely join others in deals profitable to ourselves, usually without worrying about their ultimate morality. Monetary arrangements of this sort, now an essential part of our economy, were once condemned as usury by the Christian church. In application to human behavior, Darwin (1871, chapter 5) called reciprocity a "low motive."

Reciprocity in nature is strictly limited by the necessity of safeguards against cheating. No matter how great the collective benefit might be from a cooperative venture, the necessary behavior will not evolve if the benefits can be enjoyed by freeloaders. It might be that a gregarious species could reduce its rate of loss to predators by having each individual warn the others with a loud call when it sighted an enemy. This would in no way assure that the calling would evolve. If the call attracted the notice of the predator to the caller it might increase its likelihood of being the victim. This cost would not be borne by an individual that merely sought its own safety on seeing a predator. The silent freeloader would survive better than the caller so that the sounding of the alarm would not be stable in evolution. If such behavior miraculously evolved in the population, favorable selection for freeloading would soon suppress the calling behavior and remove the collective benefit.

This sort of reasoning has led students of animal behavior to favor kin selection over reciprocity as an explanation for alarm calls, and predictions based on the kin-selection model have been borne out in recent studies. Alarm calls of various sciurid rodents are given mainly by those age-sex categories most likely to have close kin nearby, and the actual presence of individuals with a coefficient of relationship of .25 or more (grandchildren, nephews, aunts) increases the likelihood of calling (Dunford 1977; Sherman 1980). Animals sound alarms to warn their close relatives. If such warnings also benefit the group as a whole, that is an incidental consequence of selection for nepotism, and of no direct bearing on the evolution of alarm calls in gregarious animals. It could well work the other way around, with alarm calls important in the evolution of gregariousness. If my neighbors shout at their children when they see a bear approaching, it behooves me to stay close enough to hear their alarm calls. The more such neighbors there are, the greater my safety. I should stay with them if I can, and encourage them to stay with me.

A more important factor was formulated by Hamilton (1971) and later called the *dilution effect* by Terry Lim in an address to the Society for the Study of Evolution in 1980 (see also Wolf 1983). The effect comes simply from an advantage in maximizing competition for the bad things in life. In a room with ten malaria-carrying mosquitoes, I would feel more threatened alone than as one of a large crowd. If I were a Christian in the Flavian Amphitheater, I would want as many Christian companions as I could muster to share the attention of the lions. Perhaps the best technical term would be the

Saint Ignatius Strategy, after the bishop of Antioch who was the earliest on record to make use of this principle. Various groups of animals seem to use it more successfully than Saint Ignatius did (Stuart Altman 1974; Kaufmann 1974). Being companionable so that evil will befall a companion instead of oneself is of little use as a romantic's "exemplar for human conduct."

As a general rule, a modern biologist seeing one animal doing something to benefit another assumes either that it is manipulated by the other individual or that it is being subtly selfish. Its selfishness would always be defined in relation to its single ultimate interest, the replication of its own genes. Nothing resembling the Golden Rule or other widely preached ethical principle seems to be operating in living nature. It could scarcely be otherwise, when evolution is guided by a force that maximizes genetic selfishness.

D. S. Wilson provides a vivid account of the moral subversiveness of natural selection:

> The evolutionary ecologist derives a definition of self-interest from the concept of relative fitness. Briefly stated, this means that if a trait is to be selected, it must increase the fitness of its bearer relative to the fitness of other members of the population. An absolute increase in the bearer's fitness is not enough. If the fitness of others is increased even more, the trait will be selected against. Conversely, a trait that decreases personal fitness may be selected if it does even more damage to other members of the population. (1980, 5)

To be fair, I have to qualify my use of this passage. The current concept of fitness is indeed a ratio. An individual's ability to transmit its own genes is the numerator, and the average ability for the population is the denominator. The numerator is more readily responsive to individual action. It is safer and easier to feather one's nest than to unfeather everyone else's. For this reason *spite*, defined in biology as the active harming of others at net cost to oneself, seems to be rare. Organisms are regularly competitive but seldom really spiteful.

I should also concede that the quotation is out of context. It was part of an argument that living nature could not really be as bad as the explicit statement of the modern theory of natural selection would imply. Wilson was arguing for special forms of group selection, mainly what he calls trait-group selection, as forces that could favor adaptive organization at a more inclusive level than that of single individuals. Group selection is usually conceived as operating by the extinction and proliferation of local populations, with the result that such groups become better able to avoid extinction and spread to new localities. Most evolutionary biologists today believe that group selection must take place, but that it is a weak force, with little explanatory value for the phenomena that they encounter.

The passage from Wilson reminds me of Bernard Shaw's reaction to Darwinism and reason for rejecting it: "It seems simple, because you do not at first realize all that it involves. But when its whole significance dawns upon you, your heart sinks into a heap of sand within you. There is a hideous fatalism about it, a ghastly and damnable reduction of beauty and intelligence, of strength and purpose, of honor and aspiration" (1921, 33–34).

Even if it turns out that group selection is more potent than is admitted by current orthodoxy, it would make little difference for a moral evaluation of Huxley's cosmic process. If some populations are consistently better than others at maintaining themselves and giving rise to new populations, they achieve their success by causing the failure of less favored groups. To claim that this is morally superior to natural selection at the level of competing individuals would imply, in its human application, that systematic genocide is morally superior to random murder. This conclusion is forcefully recognized in Alexander's (1987) dictum that the suppression of selfishness between groups is much more urgent than its suppression between individuals. As Shaw saw so clearly, there is no level of inclusiveness of selected entities at which the survival of the fittest is morally acceptable. The morally acceptable goal in relation to survival has to be "the fitting of as many as possible to survive" (p. 82). So I conclude that natural selection really is as bad as it seems and that, as Huxley maintained (p. 83), it should be neither run from nor emulated, but rather combated.

Examples of the Triumph of Selfishness

The theory of natural selection identifies the self-interest of every individual organism as the maximal representation of its own genes in future generations. There is no encouragement for any belief that an organism can be designed for any purpose other than the most effective pursuit of this self-interest. My discussion of this theory has referred mainly to activities that do not seem selfish, like sharing food or sounding an alarm that helps others to avoid a predator. Even the victim of manipulation might be said to show the often-praised virtues of humility and charity and of turning the other cheek. Now I will examine more obvious examples of the pursuit of self-interest as practiced in the biological cosmos. I will need no theoretical subtleties to show their gross selfishness and moral unacceptability. My intent is to summarize a representative selection of phenomena in a straightforward fashion, not put on a horror show. More clinically detailed or melodramatic accounts are available (Gargett 1978; Dillard 1974). Occasionally the emotional trauma of observing some routine destruction sneaks into the technical literature, as in the fragment of field notes on ground squirrels published by Sherman (1981) and Blaffer Hrdy's (1977, 76; 1979) admission that her own tears were among the problems to be overcome in pursuit of the "high drama" of langur family life.

Kin selection does not assure that relations between relatives will be friendly, nor need a mutually advantageous coalition, like that between mates, be really amicable. Trivers was the first to see clearly the normalcy of confict within the family. His analysis of parent-offspring conflict (1974) showed that evolution favors offspring that try to get more than their fair share of resources from parents. Any success in this attempt is achieved at a cost to parents and other offspring. The same evolutionary process favors parents that try for maximal reproductive success per unit of expenditure. They can achieve this by a precise optimum compromise between numbers of offspring and benefits to each. If the parents succeed it means that each young gets less than its own optimum allotment of resources.

Trivers discussed this conflict mainly for weaning. There comes a time in the development of a litter of kittens when weaning is in the best interests of the mother, but continued nursing is best for each kitten. The kitten ideal is to continue nursing but not have its litter mates do so. The coefficient of relationship between any two individuals in this family is 0.5, not 1.0, and this means partly different genetic interests and different individual optima. A result is the noisy weaning conflict often observed in mammalian family life. Even in primates and other mammals that normally have one young at a time the weaning conflict can be intense (Jeanne Altman 1980). Successive young are ultimately in conflict with each other, just like litter mates, because continued nursing may delay the next pregnancy or reduce nutritive reserve on which the mother depends for survival to the next breeding season. Only mammals can have a weaning conflict, but analogous strife is found in all groups in which parents provide services for developing offspring. Burger (1981) describes the lively dissention over termination of parental feeding of young in herring gull colonies. Conflict within the family can be severe even in the proverbially cooperative social insects. Efforts by worker ants to lay eggs is sometimes successful and sometimes the cause of destructive violence (Bourke 1988).

Conflict between mates or potential mates is complicated and often bellicose. Many recent accounts of reproductive behavior in wild animals are tales of sexual intrigue full of deception, desertion, double-dealing, and sometimes lethal violence. Conflict arises, as Trivers (1972) explained in detail, because of basic differences in reproductive physiology of males and females in most species. A female may need to mate with a male to produce offspring, but a single mating may be enough to fertilize all the eggs she has or all she can produce in that breeding season. Mating once with the best male available can be a better strategy than mating once with the best and once with the second best. Such a female can be expected to try to optimize her choice of male, as well as the timing, locality, and all other circumstances that might influence the ultimate success of the eggs to be fertilized. For the male, reproductive success may be largely a matter of how many females he can inseminate. He can be expected to make use of every available opportunity to fertilize eggs and to seek out or try to produce such opportunities. The common outcome is for males in a population to spend much time and effort making mating attempts while females just as persistently avoid them. A tank full of guppies is a scene of constant conflict between males and females. Such is the power of romantic fancy that many people find such displays attractive. Daly (1978) reviewed the phenomena of mating conflict and argued that it exacts a high cost for survival and well-being in many animal populations.

Incest can be a special arena of conflict between male and female. Parker (1979) showed that with any decrease in offspring fitness with closer kinship between parents, there must be a range of kinship within which matings would be advantageous for a male but disadvantageous for a female. She would get offspring of reduced fitness instead of others of normal fitness. He would get offspring of reduced fitness in addition to whatever other offspring he has. For her, any reduction in offspring fitness is a net disadvantage. For him, kin selection would lead to favorable selection of incest avoidance only if the loss to his sister's fitness exceeded a certain threshold. Undoubtedly, it is often exceeded so that even males may actively avoid incest. Huxley seems to have assumed that incest would be normal in human society except for the sex urge being "profoundly modified by training" (p. 116, n. 23). The modern view is essentially that proposed by Edward Westermarck in 1891 (Daly and Wilson 1983, 307). Each individual trains itself by negative imprinting on close associates in early childhood, so the urge to avoid incest is instinctive, but the individuals to be avoided as mates must be learned.

The possibility of cuckoldry adds another dimension to male-female conflict for species in which fertilization is internal and males contribute to the care of offspring. A female songbird may be fertilized, willingly or unwillingly, by a neighbor rather than by her ostensible mate, or her mate may die or desert. Her best strategy is then to pretend that she has a full complement of unfertilized eggs for her mate, or to allow herself to be courted and won by another male. He could still sire a fraction of her clutch, but would waste at least part of his investment on another male's offspring and raise fewer young of his own than he otherwise might. The result is an evolutionary arms race between a female's ability to deceive and a male's ability to detect deception. Many details of courtship in birds have recently been interpreted as a dialogue based on male suspicion and female representations of fidelity. A male's effort to sequester and monitor his mate for a period of time that minimizes the possibility of cuckoldry may be an essential part of courtship. Evidence for such interpretations is provided by swallows (Beecher and Beecher 1979), bluebirds (Gowaty 1981; Power and Doner 1980), and doves (Lumpkin et al. 1982). The courtship period for females provides a basis for subtle decisions on choice of mate or territory, and on whether it might be better to pair monogamously with a male on an inferior territory or to share a male with another female on a better one (Verner 1964; Wolf and Stiles 1970).

There may be benefits to a female in undetected adultery with one or more of her mate's rivals. A female mouse can thereby make the rival less likely to kill her young later on (Blaffer Hrdy and Hausfater 1984; Elwood and Ostermeyer 1984). Adultery can also be a safeguard against a mate's sterility. If a male bird has a ten percent probability of being partly or entirely sterile, his mate has a ten percent probability of losing all or part of her brood if she is strictly monogamous. Insemination by two males would reduce the hazard to one percent. That females often copulate with males other than their mates has been shown by surgically sterilizing the males. About half of the mates of sterilized redwing blackbirds studied by Roberts and Kennelly (1980) produced fertilized eggs. A genetically determined color variation enabled Lank et al (1988) to calculate an cuckoldry rate of about three percent for the snow goose.

These observations need not indicate adaptive bet-hedging by females. Actual or suspected adultery may expose a female to rejection or possibly violent attack by her mate (Barash 1976), and it may be that much of the measured illegitimacy results from rape. Even a brief absence of her mate may make a female vulnerable to rape by neighboring males in the goose colony studied by Mineau and Cooke (1979). An unguarded female mallard may be raped so persistently by gangs of males that she drowns (Barash 1977). Cheng, Burns, and McKinney (1982) showed that rape is a normal part of mallard reproductive behavior. The most likely victims are those for which rape is most likely to result in fertilization. Burger and Beer (1975) observed 162 males that succeeded in mounting unwilling females in a gull colony, and numerous unsuccessful rape attempts. In some insects it would appear that females mate only as a result of male violence (Parker 1979), and in some sharks it appears that mating takes place only after injurious attack by males

(Pratt 1979). Rape occurs in turtles (Berry and Shine 1980) and newts (Verrell 1982), and it can be homosexual in a parasitic worm. By inseminating rivals a male may make it difficult for them to inseminate females with their own sperm (Adele and Gilchrist 1977). The concept of rape would most clearly include mating as a result of threat or actual violence by males. A more theoretically useful definition would include any circumvention of mate-choice mechanisms used by females. By this definition it would be of common occurrence in plants (Janzen 1977; Willson and Burley 1983). In animals, it should include the use of deception and stealth. A subordinate male frog may hide near a calling dominant so that a female attracted to the caller may be intercepted and mounted and have her eggs fertilized by the subordinate (Howard 1978; Perrill, Gerhardt, and Daniel 1978; Forester and Lykens 1986). Subordinate male sunfish, sometimes with the color and behavior of a female, may lurk about the nests of dominants. They may then dart into the nest of a spawning pair, release a cloud of semen, and flee any attack by the dominant male (Dominey 1980; Gross 1979; Arak 1984). Genetic studies now in progress may determine the extent to which this stealth and female mimicry actually succeed. Similar phenomena occur in sticklebacks (Rowland 1979) and, in modified form, in species with internal fertilization. For example, Farr (1980) showed that by a combination of stealth and speed a male guppy may sometimes circumvent attempts at rejection by females. Even in species in which mating does not normally occur without female consent, an incapacitated female may be quickly inseminated, perhaps by a functionally inappropriate male. Fish hybrids can be produced routinely by presenting males of one species with anesthetized females of another (Bowden 1969). Male fruit flies mate more readily with anesthetized females than with active ones (Ehrman 1964).

Besides adultery and rape, just about every other kind of sexual behavior that has been regarded as sinful or unethical can be found abundantly in nature. Brother-sister matings are the rule in many species (Hamilton 1967). Masturbation is common in mammals (Beach 1964). Hershkovitz (1977) noted that it was especially likely for juvenile male marmosets whose mothers are in heat. Males may mount obviously pregnant females or other inappropriate objects

(Stuart Altman 1962). Homosexual behavior is common in a wide variety of vertebrates (Beach 1978; Organ and Organ 1968) and there may be some examples of homosexual preference (Weinrich 1983).

The killing of other members of the same species is a frequent phenomenon in a wide variety of forms and contexts. Simple cannibalism is the commonest form of killing, and Polis's (1981) review indicates that it can be expected in all animals except strict vegetarians. It is a general rule among fishes. Aquarium keepers find that speedy separation of young from adults is necessary to prevent consumption of the young in most species. The phenomenon is widespread in nature, with an extreme example provided by the walleye. Cuff (1980) found that stomachs of large walleyes contained smaller ones, which had eaten still smaller ones, for at least a fourfold cycle of cannibals within cannibals.

Everyone knows the story of a female spider killing and eating her mate; this behavior has been documented for about thirty diverse species of spiders and insects (Buskirk, Frolich, and Ross 1984). Other sorts of cannibalism among insects are also common (O'Neill and Evans 1981). A male wasp searching for a female may be stung and taken home as food for young by a female already mated. Sherman (1981) showed that about eight percent of the young in a colony of ground squirrels are killed by members of their own species. A male may raid a nest to kill and eat one of the young. A female may raid the nest of a competitor and kill all the young (but not eat them).

Conspecific destruction in fishes can also take special forms, several of which are shown by the mottled sculpin (Downhower and Brown 1981). Larger males are better egg guarders than small ones, and females prefer them, but if there is too great a difference in size he may find her more tempting as a meal than as a mate. Egg guarding is a stress for males, because it does not allow for productive foraging. The fast is genetically justified only if it results in a certain minimum of hatched young. If the male gets only a single clutch of eggs and no more for a few days he may eat them and abandon the attempt to reproduce that year. If he gets a succession of clutches from different females, and is sufficiently hungry after a long bout of egg guarding, he may eat the last batch instead of waiting for them to hatch. The eating of eggs or young in other individuals' nests is common in many fishes (DeMartini 1987; Keenleyside 1972 Rohwer 1978) and amphibians (Crump 1983).

Cannibalism specifically directed at one's own young is also practiced by insects (Eichwort 1973) and by sticklebacks, which also destroy eggs and nests of neighbors (Rohwer 1978). Destruction of the young of rivals can be a major source of mortality in birds (Parsons 1971; Picman 1977; Trail, Strahl and Brown 1981). Destruction of one of a pair of young, with consumption of the victim, is frequent in birds of prey (Stinson 1979); this killing may be by a brother or sister. Gargett (1978) recorded hundreds of attacks by a larger black eagle chick against a smaller nest mate over a period of three days, when the smaller finally died from its wounds and starvation.

Blaffer Hrdy (1977a, 1977b), Hausfater and Blaffer Hrdy (1984), Hoogland (1985), and Sherman (1981) have documented the previously unappreciated frequency of various forms of infanticide among mammals, including many primates. Among rodents and carnivores the killer often eats the victim. It is now generally realized that infanticide rates of many human populations were (or are) sufficient to have important demographic consequences (Bugos and McCarthy 1984; Dickeman 1979; Langer 1974; Kellam 1974; Polansky, Hally, and Polansky 1975).

Mammalian infanticide may evolve as a male adaptation to female reproductive physiology, because lactation often inhibits ovulation. For a male's reproduction, the essential resource is an ovulating female. A female deferring ovulation while nursing another male's young is only a potential resource. He may be able to hasten the change from potential to actual by killing the unweaned infant. For females, such young are of highest importance, because with a bit more investment they can be turned into self-feeding juveniles, so the death of the young would be a major loss for the female, but only a minor gain for the male. Unfortunately, such issues are settled not by any principles of justice or collective cost-benefit considerations, but strictly on a might-makes-right basis.

Blaffer Hrdy (1977a, 1977b) described dramatic examples of this sort of male-female conflict for a monkey that often lives in groups of related adult females, their young, and one temporarily dominant male who enters the group from the outside. Males keep such harems only as long as they can avoid being deposed by rivals. Sooner or later, some rival will succeed in defeating the currently privileged male and taking over his mates. When this happens, there is an immediate conflict between the new male and any female with an unweaned infant. The sooner the infant stops nursing, the sooner she will turn into a valuable resource for the new male, and the quickest way to get it to stop nursing is to kill it. This is not always an easy task. The male is bigger and stronger than any female, and can kill an infant with an efficient bite into the skull, but he may fail despite repeated attempts. A female's motivation to protect her infant may be greater than his desire to kill it, and it may be necessary for him to fight both the mother and one or more other relatives of the infant. Members of the harem are closely related, and a grandmother or aunt or older half sister has half as much at stake in the infant's survival as the mother has. Kin selection leads to the formation of coalitions of related females in opposition to infanticidal males.

In the absence of ideology, such an alliance can result only from kin selection, and never from anything analogous to feminism or a sense of justice or group welfare. If a mother loses her infant, she quickly comes into estrus and accepts her infant's killer as the father of her next offspring. If infanticide by a male raises his fitness even slightly, it is in a mother's interest to have her son practice it if and when he succeeds in taking a harem from some other mother's son.

Many conspecific killings result from violent contests over resources, a common event among insects (Parker 1979). The disputed resource is perhaps most often a reproductive female. Male smelt are often injured or stranded in struggles for access to females (Fridgeirsson 1976). In large mammals with horns or antlers, death or debilitating injury from fighting over females may claim five to ten percent of adult males every breeding season (Clutton-Brock et al. 1979; Wilkinson and Shank 1976). It was noted by E. O. Wilson (1975) that death from strife among neighbors tends to be recorded for any wild animal population carefully observed for a thousand hours or more. Consider for comparison the annual homicide rate of .0003 for Houston, recently the most murderous of major American cities (*World Almanac and Book of Facts 1982*). It would be necessary to keep ten Houston residents under continuous observation for three centuries to make it likely that one murder would be seen.

Males of our own and most other vertebrate species are more likely than females to kill or injure others of their own kind (Daly and Wilson 1988), but competition among females may not be so much of lesser intensity as of greater subtlety. Females commonly deprive each other of resources in various ways, aggressively interfere with one another's reproduction, attack one another's young, and actively aggravate male competition when it serves their interests (see Silk et al. 1981, and several contributions in Wasser 1983).

Losses from sexual conflict may go far beyond direct results to participants. Fighting males may accidentally injure or kill females or young in such diverse animals as seals (LeBoeuf 1974) and dung flies (Parker 1979). Bright colors useful in attracting females or intimidating rivals may attract predators. This is certainly true in many fishes (Endler 1978; Farr 1975; McPhail 1969) and insects (Burk 1981). Elaborate sexual and competitive behaviors also increase vulnerability either by their conspicuousness or the distractions of participation. Schaller (1972, 243) found that predation by lions was especially high on fighting warthogs or courting reedbucks. Tuttle and Ryan (1981) noted that a predatory bat used frogs' mating calls to locate prey, and Belwood and Morris (1987) found that calling male katydids may attract insectivorous bats. Other losses are less dramatic but still real. In many species a large size is advantageous for males in winning females, but is deleterious in other respects. Baker and Fox (1978) found that larger male grackles were especially vulnerable to a chemical eradication technique.

Such observations support the common assumption that females more closely approximate the engineering optima for a species but males compromise these optima because of the requirements of sexual competition. Anatomical changes may go far beyond scaling for larger size. Antlers and horns are nutritionally expensive structures, and males that use them in combat show major structural modifications related to the offensive use of these weapons and to withstanding violent collisions with rivals (Schaffer 1968). The trade-offs evolved for effectiveness in sexual competition must require compromises with many other adaptations. The consistently higher human male rates of mortality and morbidity measure the price of features useful in sexual rivalry (Trivers 1985, chapter 12). Lande's (1980) calculations support the recently common assumption that increased effectiveness in sexual competition increases the likelihood of extinction.

Many of the unpleasant phenomena reviewed above have been described only recently. They are likely to be impressive only if unexpected, as a result of the blindness of romanticism. None of them are really needed for the argument being advanced; the inescapable arithmetic of predation and parasitism should be enough to show that nature is morally unacceptable. This should be so even if some of the romantic fictions were really true, which they are not. It is sometimes claimed, for example, that predators take only what they need and avoid unnecessary killing. The excess carnage that results from attacks by bluefish or swordfish on their prey are a well-known example of wasteful predation (Bigelow and Schroeder 1953). Foxes raiding gull colonies often kill far more prey then they eat (Kruuk 1976). Hunting by primitive tribes, for instance, the driving of herds of large ungulates off cliffs (Speth 1983), was often grossly wasteful. Sih (1980) argues the general importance of wasteful predation, especially the eating of only choice parts when prey are abundant enough to permit such extravagance.

The survival of one organism is possible only at great cost to others. The moral message in this obvious fact has been recognized by many philosophers and humanists, despite the general prevalence of romanticism. Tennyson confessed to a confusion and pessimism about Nature when

> considering everywhere
> Her secret meaning in her deeds
> And finding that of fifty seeds
> She often brings but one to bear.
>
> *In Memoriam* 55

He must have realized that one-in-fifty would be extraordinarily favorable odds for all but a small minority of the world's species. More recently Dillard deplored "a world in which half the creatures are running from—or limping from—the other half." She described books on parasite life cycles as "hellish hagiography . . . the devil's summa theologica" (1974, 229). Of animals killed by predators and parasites she says "Wonderful things, wasted. It's a wretched system" (ibid., 175). Each of Tennyson's forty-nine lost seeds should also count as one of Dillard's wasted wonders. Each could potentially become a complete adult plant, and even if we discount the possibility of a plant having any perceived value to itself, it can hold great esthetic appeal to any being with an esthetic sense. That appeal could only increase with a biologist's awareness of the plant's superb engineering.

I will end this section with a thought experiment to give final emphasis to the world's wretchedness that Dillard deplored. Imagine a device of great beauty and utility and precision, perhaps an artfully designed watch, and imagine that a large number of such watches are laid out on display. Each is unique, with slight differences from any other in appearance and working details, but the differences are all harmoniously compensating so that every watch is reliable and accurate. Someone now approaches and, to secure some minor gain, smashes each of the watches with a hammer and continues the destruction until every part is destroyed beyond recovery. For me, a personally attractive wielder of the hammer would make the deed worse, not better. If the act by the hammerer were compelled by some cosmic process, it would not absolve the blame, but only transfer it.

Now replace each watch with something far more worthy of admiration, a krill (*Euphausia superba*) of the Antarctic seas. Each adult is a beautifully streamlined animal of shades of red, with optically precise photophores on eyestalks and abdomen. It has developed against enormous odds for many months, through a complex history of larval and juvenile forms to its present state. Merely the mechanical system of appendages and mouth parts by which it feeds (Hamner et al. 1983) would compare favorably in engineering to the watch. Unlike the watch, the food-gathering machinery is constantly subject to corrective control by a nervous system.

On comes a blue whale, to me an extremely appealing animal. In a few minutes the baleen plates of its mouth net hundreds of thousands of krill. Then the tongue occludes the opening, the jaws close, and a force of many tons ends hundreds of thousands of lives. The rest of the alimentary canal of the whale completes the destruction. The gain for the whale is minor, and to get any significant nutrition it must repeat the process many time a day. The life history of any one whale is merely a microscopic detail in the carnage of the seas.

PROSPECTS FOR MORALITY IN AN IMMORAL WORLD

Huxley viewed the cosmic process as an enemy that must be combated. I take a similar but more extreme position, based both on the more extreme contemporary view of natural selection as a process for maximizing selfishness, and on the longer list of vices now assignable to the enemy. If this enemy is worse than Huxley thought, there is a more urgent need for biological understanding. As Singer, a philosopher who has taken the trouble to learn modern evolutionary theory, put it in precisely this context, "the more you know about your opponent, the better your chances of winning" (1981, 168). Inadequate knowledge is likely to lead, as Symons (1979, 65) noted specifically for problems of human sexuality, to counterproductive tactics.

The modern view also gives added emphasis to Huxley's "apparent paradox that ethical nature, while born or cosmic nature, is necessarily at enmity with its parent" (1893–94, 9:viii). How could maximizing selfishness produce an organism capable of often advocating, and occasionally practicing, charity towards strangers and even towards animals? Huxley dealt only sketchily with this problem, but anticipated (p. 80) that biology would someday "arrive at an understanding of the aesthetic faculty" that makes people want a future society that "demands self-restraint; in place of thrusting aside, or treading down, all competitors, it requires that the individual shall not merely respect, but shall help his fellows; its influence is directed, not so much to the survival of the fittest, as to the fitting of as many as possible to survive" (p. 82).

A number of recent biologists have concerned themselves with this problem (Alexander 1979, 1987; Barash 1982; E. O. Wilson 1978). Not surprisingly, they all invoke kin-selected altruism and adaptations for taking maximum advantage of reciprocation. They attempt to show why such factors operating in Stone Age society should produce attitudes that favor the development, in modern societies, of broadly inclusive ethical systems. A similar attempt is made by Singer (1981), who proposes that the human capacity for reasoned argument constantly imposes a need for the public justification of personal action. As one's actions affect a broader circle of individuals, a broader justification is required. Thus was Plato led to champion the welfare of all Greeks, not merely of all Athenians (ibid., 117). This now seems a comically parochial attitude, but in the tiny world that Plato inhabited it may have seemed radically idealistic.

As Singer's ever expanding circle of consideration reaches individuals of negligible relationship or likelihood of reciprocation, it puts an altruist into ever more basic conflict with Huxley's cosmic process. This combat, between the "microscopic atom" and the "illimitable macrocosm" (p. 59) may seem a bit one-sided. Isn't Dillard showing a pathetic megalomania when she exclaims, "I came from the world, I crawled out of a sea of amino acids, and now I must whirl around and shake my fist at that sea and cry shame!" (1974, 177)? Are Huxley and Dillard perhaps like Job's neighbors urging him to rebel against oppression by an arbitrarily malicious, all-powerful, and omniscient god?

No, not omniscient. The evolutionary process is immensely powerful and oppressive, but, unlike Job's God, it is abysmally stupid. It can reliably maximize current selfishness at the level of the gene, but it is blind to future macroscopic consequences of current action. It does not have the sense to realize that mechanisms evolved for practicing unfair nepotism or making self-seeking deals with others can be subverted in the interests of broad altruism. Thus can Huxley's "ethical nature" (p. 75) emerge incidentally from biological nature. It is one of many examples of evolutionary changes with important future consequences entirely unrelated to the selfishness responsible for the changes.

I will mention two more of these examples. Photosynthesis was evolved by primitive bacteria as a way of acquiring energy-rich compounds for anaerobic metabolism. Free oxygen was a noxious byproduct tolerable only in minute traces. Concentrations were kept tolerable by rapid loss to an anoxic environment. As an ever worsening shortage of organic compounds led to increasing reliance on photosynthesis, the oxygen concentrations in some habitats approached critical levels. This primeval environmental crisis was met, in at least one lineage of organisms, by the evolution of biochemical machinery for detoxifying oxygen. The resulting tolerance of oxygen was a prerequisite to a later evolution of mechanisms for putting it to positive use in aerobic metabolism, which allows much higher levels of energy use. Ultimately, human life was made possible because some bacteria developed a way of getting more than their fair share of organic matter.

My second example is the much more recent origin of the human hand. Its subtle and versatile engineering was fully appreciated by Galen about eighteen hundred years ago (May 1968). Over the last million years it has apparently been perfected by selection for precise manipulation of objects, but such selection could only be brought to bear on an organ already adapted to grasping and rotation. This earlier stage was produced by selection for arborial locomotion, especially brachiation, which demands the ability to turn the body with a hand holding an overhead limb. The production of an organ capable of threading needles and turning doorknobs was made possible as an incidental consequence of selection to be better than one's neighbor at swinging through the trees. The helping hand of the good Samaritan and the motivation for its use raise no question on the malice or power of natural selection. They merely show that this persistent and powerful enemy is a mindless fool.

This is the one real advantage enjoyed by those who would follow Huxley's banner into combat. It is hardly a cause for complacency. The enemy is indeed powerful and persistent, and we need all the help we can get in trying to overcome billions of years of selection for selfishness. If biology can help us to understand the enemy, it should be of some help. Singer, as noted above, has pointed out that it would be unwise to rely uncritically on attitudes arising from family life and group loyalties; they are based on narrow genetic self-interest. Attitudes that arise directly from biology must exist to serve the enemy's cause, not ours. Huxley (pp. 27, 44), and more recently Campbell (1975) and Power (1981; 17), noted a parallel between the effects of natural selection and the doctrine of original sin as explanation for human misconduct.

The unparalleled human capability for symbolic communication has an incidental consequence of special importance for ethics. In biological usage, communication is nearly synonymous with attempted manipulation. It is a low-cost way of getting someone else to behave in a way favorable to oneself. To be successfully manipulated may or may not be adaptive. When a jay heeds the message "I am toxic" in the color pattern of a monarch butterfly, it is benefited by this manipulation. When it gives the same reaction to the same message from a viceroy, it is deprived of nourishment.

Human speech is so effective a form of communication that once evolved it gave rise to a system of information transfer to rival the transmission of genes in reproduction. It made elaborate teaching possible, and it enormously augmented a purely cultural evolutionary process. The mechanism of cultural evolution is fundamentally different from that of genetic transmission, but the resulting evolutionary processes have many formal aspects in common. Like genetic transmission, progress of cultural information by verbal messages or direct imitation can be tracked by phenotypic effects on recipient individuals. Recognizable cultural elements—I prefer Dawkins's term *memes*—can wax or wane in a population, and such quantitative effects can be treated by mathematical devices not greatly different from those of population genetics (Cavalli-Sforza and Feldman 1981; Mundinger 1980; Lumsden and Wilson 1981). A meme (e.g., a new song, a religious cult, a food fad) will spread in a society if that society constitutes a favorable environment for its spread, in an obvious analogy with selection acting on a gene. Perhaps the natural selection of memes is a bit less deplorable than that of genes, for human beings are not as stupid as nature. They can sometimes envision, and evaluate in advance, the social effects of adopted a meme.

But ultimately the ethical problems caused by genes and memes are basically the same. The one necessary and sufficient reason for a meme spreading through a human population is that it is good at getting itself spread. It is not necessary that it enhance the biological fitness or perceived well-being of its practitioners or that it increase general prosperity. I am sure that cigarette smoking has spread more rapidly through some populations in recent decades than have many sound public health practices. Dawkins (1976) identifies the natural selection of cultural features as another kind of attack by the tenacious and powerful enemy.

In its recognition of the evolutionary importance of manipulation and the added manipulative potential of language, biology can add its voice to the chorus of warnings on the dangers of propaganda in the mass media. Language, no doubt, did mischief enough in primitive societies. Kinship and reciprocity may have made communication adaptive for recipients most of the time, but there must have been numerous exceptions. It may often have been adaptive for the patriarch to say to the youth, "You must marry this individual rather than that", or "We (you) must destroy the Evil Ones to redeem our (my) honor." For this reason, an appreciable level of suspicion and skepticism would be adaptive for the youth. Uncritical acceptance of such messages could well be contrary to the health and happiness of the recipient, and contrary to most broadly conceived ethical systems.

Reasoned analysis prompted by suspicion is a fortunate human tendency. Throughout human history it has undoubtedly aided the receivers of messages in deciding (1) whether they gain or lose by complying, and, (2) if they are likely to lose, whether defiance will bring a retaliation worse than the loss. These can be problems of great subtlety, with immediate and long-range aspects, and with consequences for kin and valued associates. The ability to deal with such questions was designed by natural selection to spread selfish genes. In its boundless stupidity, this evolutionary process incidentally designed machinery capable of answering other sorts of questions, such as, Is the message one of help or harm for what I really want for the world?

Answers to this sort of question might well serve morality rather than selfishness. Machinery designed for competitive self-seeking in the tribal microcosm can serve altruism in the global macrocosm. We can use abilities developed for petty intrigue to deal with sermons coming from the pulpit, printed tracts by sociobiologists, or manipulative arguments from politicians on television screens. People can now espouse remote and inclusive ideals far removed from the selfishness that gave rise to the power to do so. It was inevitable that people in the novel civic environments of the last few millenia would develop aspirations for such things as the dictatorship of the proletariat, or the triumph of the master race, or the saving of souls. Because such strivings are beyond the direct action of natural selection, I have some hope that some such cause can provide the humane artifice that can save humanity from human nature.

A sociobiological awareness ought to help in evaluating courses of action in relation to whatever one chooses as an ultimately worthy cause. If a proposal causes an immediately favorable reaction, the awareness should prompt the question, Can I reasonably maintain that what is proposed will have a beneficial effect, or is my reaction favorable merely because, in a normal environment, it would be favorable to the survival of my genes? Singer puts it nicely when he says, "Discovering biological origins for our intuitions should make us skeptical about thinking of them as self-evident moral axioms. . . . Far from justifying principles that are shown to be "natural," a biological explanation is often a way of debunking the lofty status of what seemed a self-evident moral law" (1981, 70–71). The sociobiological imperative is thus a negative one: beware of manipulation by selfish individuals, or selfish institutions, or our own selfish genes.

We should be especially suspicious of indoctrination that appeals directly to such biological factors as nepotism, as when the more powerful in a group tell the less powerful that members of their group are brothers, and thereby worthy of special consideration (Johnson, Ratwik, and Sawyer 1988). Would the special consideration really help these so-called brothers, or merely those from whom the message is coming, and what are the implied limits to the brotherhood?

Modern biology also helps us to identify ourselves as possible soldiers in Huxley's army and to clarify the distinction between ourselves and the enemy. Each of us at conception got a unique genotype that never existed before. Unless there is an early division of the embryo, it will never be duplicated. This genotype has no significance in evolution beyond its brief and minor influence on rates of increase and decrease of component genes. It is of the utmost personal significance because it directs individual development and controls vital functions. In these processes each gene interacts in various ways with the rest of the genotype and with environmental conditions. On the average over many generations, a gene must produce results favorable to its own replication, but the particular interactions in our own development were unique and unpredictable and may deviate markedly from average effects. We are all special genetically, besides having unique individual histories that give us special collections of memories and attitudes that constitute the self. There is no justification for any personal concern with the long-term average proliferation of the genes we got in the lottery of meiosis and fertilization. As Huxley was the first to recognize, there is every reason to oppose any tendency to serve such interests, "to refuse any longer to be the instruments of the evolutionary process" (p. 63). In Dawkins' more modern terms, we must "rebel against the tyranny of the selfish replicators" (1976, 215).

Stent objects that Dawkins' concept of rebelling against one's own genes is a "biological absurdity . . . no more possible than the natural selection of the unfit" (1978, 19-20). I can think of no more fitting response than Huxley's to this same challenge: "If the conclusion that [the natural and the ethical] are antagonistic is logically absurd, I am sorry for logic, because, as we have seen, the fact is so" (p. 12). It seems strange that Stent missed the relevance of whole technologies, such as hair dyeing and cosmetic surgery, based on attempts of individuals to correct perceived flaws in development controlled by their own genotypes. The combat urged by Huxley and refined by Dawkins is against the much less personal effects of our genes in many individuals over many generations. It is what Lopreato meant when he proposed that the goal of morality should be the "ultimate negation of the commandment of natural selection" (1981, 124). This is very much in the spirit of Huxley's opposition to the cosmic process.

Like Huxley and everyone else, I have my opinions on what the world ought to be like and on the best ways of moving towards that goal. Huxley did not write his essay to champion a particular view of utopia, and my purpose is merely to update his message and to characterize the enemy more clearly than was possible in his time. The updated policy for the betterment of the human condition is an attack on both the natural enemy and any institutional enemies favored by cultural evolution. In the nuclear age far more than at any prior time, we can survive only if our morality "repudiates the gladiatorial theory of existence" (p. 82) and if "in virtue of his intelligence, the dwarf bends the Titan to his will" (p. 84).

Appendix:
The History of
Evolution and Ethics

THE ROMANES LECTURE

The Romanes Lectures were established in 1892 at Oxford University by Professor George Romanes, who chose as his first two lecturers William Gladstone and T. H. Huxley. Although Romanes stipulated that the subjects of politics and religion should not be broached, neither Gladstone nor Huxley quite managed to avoid them. Indeed, given the conditions of his lecture series, Romanes's choice of lecturers was provocative, for Gladstone and Huxley had for a good part of the previous decade been in controversy in the pages of *The Nineteenth Century* over the naturalistic basis of biblical miracles such as the Creation account—which Gladstone defended with grand condescension and Huxley attacked with withering sarcasm. Gladstone had just published his *Impregnable Rock of Holy Scripture* (1890), and Huxley had published *Essays on Some Controverted Questions* (1892), which included "Mr. Gladstone and Genesis," "The Evolution of Theology: an Anthropological Study," and "Illustrations of Mr. Gladstone's Controversial Methods," among other anti-Gladstonian essays. Hence, the Oxford stage was set, so to speak, with two famous if failing antagonists, who had made the defense of orthodoxy and scientific naturalism their respective final objects in life.

Gladstone's lecture, titled "An Academic Sketch," was delivered in October of 1892 and published as a pamphlet the same year (Gladstone 1892). Sketching the history of Oxford and Cambridge Universities, Gladstone, very much the familiar Oxonian at home, if not on entirely sympathetic political territory, argued that the English universities should promote their traditional alliance with religion and retain theology as one of the fundamentals of modern education (ibid., 43). Contemplating the eventual decline of the British Empire, he reminded his audience that "throughout the range of the ancient civilisation" in the centuries of Roman decline, there was "but one conspicuous instance of a standing attempt at systematic and orderly self-government, together with the adjustment of disputes by the word rather than the sword. This example was to be found in the ordered fabric of the Christian Church, which amidst surrounding decay, the living among the dead steadily developed its organisation, and constructed its theology" (ibid., 5). Thus Gladstone, the protégé of Samuel Wilberforce, with whom Huxley had debated at the famous Oxford British Association meeting of 1860 (Leonard Huxley 1900, 1:179), once more defended Orthodoxy as the legitimate partner of education. "I should think that there is no living man," Huxley wrote to Romanes in response to Gladstone's lecture, "who, on such an occasion, could intend and contrive to say so much and so well (in form) without ever rising above the level of antiquarian gossip" (ibid., 2:351).

To his old friend John Tyndall, Huxley wrote, "Who would have thought thirty-three years ago, when the great 'Sammy' fight came off, that the next time I should speak at Oxford would be in succession to Gladstone, on 'Evolution and Ethics' as an invited lecturer?" (ibid., 2:356). Huxley was sensitive about appearing at Oxford, despite his close friendships with Jowett and Romanes himself. Fearing loss of intellectual independence—and income—he had declined the Linnacre Professorship of Physiology and discouraged preliminary offers of a Mastership of University College in 1881. Although he had received an Honorary Oxford D.C.L. in 1885, it had been after he had been turned down for the consideration years earlier at the objections of Froude.

These circumstances help to explain some of the textual peculiarities of Huxley's Romanes Lecture. In addressing his academic audience, Huxley augmented his discourse with a series of classical quotations, and drew as well on contemporary textual scholarship on Eastern religions. The footnotes come to nearly nine thousand words, three-quarters of the length of the main text. They were not, of course, articulated during the lecture itself. Huxley's quotations, largely drawn from the Greek Stoical tradition, also animated the naturalistic theme of the main argument. Moreover, by arguing that all religious traditions—Eastern, Ionian, and Hebrew alike—raise up ethical systems, Huxley countered any argument that Christianity alone produced the highly developed ethical system and sense of justice Victorians associated with civilization. Ethics, he suggested, being generic to all cultures, has a naturalistic basis in human behavior. In many respects, and quite despite Romanes's admonition to avoid politics and religion, Huxley's Romanes Lecture is the final statement of his controversy with Gladstone over the naturalistic basis of the miraculous.

THE PROLEGOMENA

Huxley wrote the "Prolegomena to 'Evolution and Ethics' " after delivering the Romanes lecture on May 18, 1893. Completed in June of 1894, the "Prolegomena" was intended as a preface to later published editions of "Evolution and Ethics," and the two are best considered as a single work. He referred to the "Prolegomena" in a letter to Frederick Macmillan as his effort "to cross the t's and dot the i's" of his Romanes Lecture (Br. Mus. Add. MSS 55210, fol. 156). The "Prolegomena," Huxley felt, concerned fundamentals that were long established in his work, but was new in the sense that it formulated a set of oppositions between artificial and natural conditions that were unique in the history of naturalistic thinking. The "Prolegomena" was addressed to criticisms of "Evolution and Ethics," concerning Huxley's apparent abandonment of a monistic, evolutionary world view. Hence, the appearance of the two essays together in volume 9 of the *Collected Essays* amounted to a revised edition of the Romanes Lecture.

To a considerable extent, humanist themes of textual analysis and scholarship occupy Huxley in the Romanes Lecture, and naturalistic speculation occupies him in the "Prolegomena." The Romanes Lecture had devoted nearly half its argument to the comparative ethical responses of ancient religions, a topic that was of considerable contemporary scholarly interest. Although the Malthusian theme is present in the Romanes Lecture, it is more elaborately explored and set apart from culture in the "Prolegomena." Hence, the "Prolegomena," less self-consciously literary in quality, is more a straight intellectual study of the naturalistic basis of Huxley's argument that humans are at strife with themselves and their conditions of existence. The "Prolegomena" develops more forcefully the dichotomy between the "state of nature" and the "state of art," between the "cosmic process" and "ethical process," as the central paradox of human existence.

<h2 style="text-align:center">EDITIONS</h2>

"Evolution and Ethics" was originally published as a pamphlet immediately after its delivery. By June 16, Frederick Macmillan had written Huxley to inform him that all but 327 copies of the original three thousand published had been sold. In republishing the Romanes Lecture in volume 9 of the *Collected Essays* in 1894, Huxley made only a few nonsubstantive revisions. He dropped the *e* in the spelling of *develope*, substituted two words for stylistic consistency, and added three footnotes—numbers 19, 21, and 23. None of these changes had any significant effect on the content of the original Romanes Lecture.

"Evolution and Ethics" and its "Prolegomena" made their first appearance together in 1894 in volume 9 of the *Collected Essays*, titled *Evolution and Ethics, and Other Essays*. It was published by Huxley's main publishers for more than thirty years, in London, Macmillan and Company, and in the United States, D. Appleton and Company. Huxley's nine volume *Collected Essays*, originally published in 1893–1894, were thereafter reprinted more than a dozen times over the following four decades. The 1893–1894 London edition of the *Collected Essays*, including volume 9, has been reprinted by G. Olms (Hildesheim, New York, 1970). A recent reprinting of American "authorized edition" of the *Collected Essays* has been issued by Greenwood Press (New York, 1968).

A reprinting of Huxley's "Prolegomena" and "Evolution and Ethics" appeared in 1947, together with three essays by Julian Huxley, Huxley's grandson. The new volume, titled *Evolution and Ethics, 1893–1943* (London: Pilot Press), joined T. H. Huxley's "Prolegomena" and Romanes Lecture with Julian Huxley's own 1943 Romanes Lecture, titled "Evolutionary Ethics." The volume began with Julian Huxley's introductory survey of the philosophical origins of ethical naturalism together with a concluding essay reflecting on post-war implications of ethical naturalism. His approach was heavily influenced by Freudian psychological theory.

This volume was published in the United States by Harper and Brothers under the title *Touchstone for Ethics: 1893–1943* (New York, 1947).

CRITICISM

T. H. Huxley's "Evolution and Ethics" and "Prolegomena" have inspired much commentary over what is now nearly the century since their original publication. Much of the commentary of Huxley's own time concerned the apparent shift in Huxley's thought. Perhaps the most extended critique was that of John Dewey (1898). Kropotkin's *Mutual Aid* (1914), although written in response to Huxley's "The Struggle for Existence in Human Society," can be taken as a Romantic view that stands opposed to Huxley's own social realism in the Romanes Lecture. St. John Mivart's review (1894) is important, not only as a theistic response to Huxley's lecture, but also as a historical reminder of Huxley's shift in viewpoint. Spencer's rebuttal in the *Atheneum* (1893) is an indispensable footnote to Huxley's psychological dualism. Andrew Seth's *Man's Place in the Cosmos* (New York: Scribner's, 1897) places the essay in the tradition of Stoical thought, linking it explicitly to Matthew Arnold's central philosophical poem *Empedocles on Etna*. E. Ray Lankester's "Nature's Insurgent Son," the Romanes Lecture of 1905, carries Huxley's ideas forward. Of special interest is Julian Huxley's commentary on Huxley's "Evolution and Ethics" in *Touchstone for Ethics: 1893–1943*, which contains Julian's reflections upon the changes of the past several decades, complete with a reconstructed teleological view of ethics. The ethical naturalism of Thomas and Julian Huxley, including the conflicts in their points of view, is examined by Steven Toulmin in *The Return to Cosmology* (1982). Interestingly, Aldous Huxley's *Brave New World* (1932), which he referred to as a "fictional essay in Utopianism," seems to have been a thought experiment with a future stabilized society of zero population growth within a controlled, collectivized industrial society, which had solved the problem of self-gratification identified by T. H. Huxley. More recent treatments of the two essays can be found in Irvine (1955), Himmelfarb (1968, 1986), Helfand (1977), Ruse (1986), and Richards (1987).

Bibliography I:
The Victorian Context

Bagehot, Walter. 1869. *Physics and Politics*. Reprint. 1956. Boston: Beacon Press.

Barrow, John D, and Frank J. Tipler. 1986. *The Anthropic Cosmological Principle*. New York: Oxford University Press.

Bartholomew, Michael. 1975. "Huxley's Defence of Darwin." *Annals of Science* 32:525–35.

Barton, Ruth. 1983. "Evolution: The Whitworth Gun in Huxley's War for the Liberation of Science from Theology." In *The Wider Domain of Evolutionary Thought*, ed. D. Oldroyd and I. Langham, 261–87. Dordrecht: Reidel.

Beer, Gillian. 1983. *Darwin's Plots: Evolutionary Narrative in Darwin, George Eliot, and Nineteenth-Century Fiction*. London: Routledge and Kegan Paul.

Bowler, Peter J. 1984. *Evolution: The History of an Idea*. Berkeley: University of California Press.

————. 1986. *Theories of Human Evolution: A Century of Debate, 1844–1944*. Baltimore: Johns Hopkins University Press.

Burrow, J. W. 1966. *Evolution and Society: A Study in Victorian Social Theory*. Cambridge: Cambridge University Press.

Carpenter, William. 1889. *Nature and Man: Essays Scientific and Philosophical*. New York: Appleton.

Clifford, W. K. 1877. "Cosmic Emotion." *Nineteenth Century* 1:411–29.

Courtney, William L. 1895. "Professor Huxley as a Philosopher." *Fortnightly Review* 64:17–22.

Darwin, Charles. 1859. *The Origin of Species . . . a Variorum Text*. Ed. Morse Peckham. 1959. Philadelphia: University of Pennsylvania Press.

————. 1871. *The Descent of Man and Selection in Relation to Sex*. Ed. John Tyler Bonner and Robert M. May. 1981. Princeton: Princeton University Press.

————. 1872. *The Expression of the Emotions in Man and Animals*. Reprint. 1955. New York: The Philosophical Library.

Desmond, Adrian. 1982. *Archetypes and Ancestors: Paleontology in Victorian London, 1850–1875*. Chicago: University of Chicago Press.

Dewey, John. 1898. "Evolution and Ethics." *Monist* 8:321–41.

Dewey, John. 1920. *Reconstruction in Philosophy*. Enlarged ed. 1948. Boston: Beacon Press.

————. 1929. *The Quest for Certainty: A Study in the Relation of Knowledge and Action*. Vol. 4 of *The Later Works of John Dewey*. Ed. Jo Ann Boydston; intro. Stephen Toulmin. 1984. Carbondale: Southern Illinois University Press.

di Gregorio, Mario A. 1984. *T. H. Huxley's Place in Natural Science*. New Haven: Yale University Press.

Drummond, Henry. 1894. *The Ascent of Man*. 7th ed. 1898. New York: James Pott and Company.

Dubois, René. 1980. *The Wooing of Earth: New Perspectives on Man's Use of Nature*. New York: Scribner's.

Engels, Friedrich. 1927. *Dialectics of Nature*. Trans. and ed. Clemens Dutt. 1940. New York: International Publishers.

"Evolution and Ethics," by T. H. Huxley. Review. 1893a. *The Atheneum*, no. 3430:119–20.

"Evolution and Ethics," by T. H. Huxley. Review. 1893b. *Oxford Magazine* 11:380–81.

Flew, Anthony. 1967. *Evolutionary Ethics*. New York: St. Martin's.

Foster, Sir Michael. 1896. Obituary Notice for Thomas Henry Huxley. *Proceedings of the Royal Society of London* 59:xlvi–lxvii.

Freud, Sigmund. 1931. *Civilization and its Discontents*. Trans. and ed. James Strachey. 1962. New York: W. W. Norton.

Galton, Francis. 1892. *Hereditary Genius: An Enquiry into its Laws and Consequences*. Ed. C. D. Darlington. 2d ed. 1962. New York: World Publishing Company.

Ghiselin, Michael T. 1969. *The Triumph of the Darwinian Method*. Reprint. 1984. Chicago: University of Chicago Press.

Gibbon, Edward. 1901–02. *The History of the Decline and Fall of the Roman Empire*. Ed. John Bury. 2d ed. 7 vols. London: Methuen.

Gillispie, Charles C. 1951. *Genesis and Geology: The Impact of Scientific Discoveries upon Religious Beliefs in the Decades before Darwin*. New York: Harper and Row.

Gladstone, William E. 1892. "An Academic Sketch." First Romanes Lecture, October 24, 1892. Oxford: Clarendon Press.

Greene, John C. 1959a. *The Death of Adam: Evolution and Its Impact on Western Thought*. Ames: Iowa State University Press.

————. 1959b. "Biology and Social Theory in the Nineteenth Century: Auguste Comte and Herbert Spenser." In *Critical Problems in the History of Science*, ed. Marshall Clagett, 419–47. Madison: University of Wisconsin Press.

―――. 1977. "Darwin as Social Evolutionist." *Journal of the History of Biology* 10:1–27.

Guizot, François Pierre Guillaume. 1885. *History of Civilization from the Fall of the Roman Empire to the French Revolution.* Trans. William Hazlitt. 2 vols. London: Macmillan.

Hanham, Harold. 1970. "Introduction." In *On Scotland and the Scotch Intellect,* xiii–xxxvii. Chicago: University of Chicago Press.

Hankins, Thomas L. 1985. *Science and the Enlightenment.* Cambridge: Cambridge University Press.

Hartley, David. 1749. *Observations on Man, His Frame, His Duty, and His Expectations.* Reprint. 1966. Gainesville: University of Florida Press.

Hearnshaw, F.J.C., ed. 1933. *The Social and Political Ideas of Some Representative Thinkers of the Victorian Age.* Reprint. 1967. New York: Barnes and Noble.

Helfand, Michael S. 1977. "T. H. Huxley's 'Evolution and Ethics': The Politics of Evolution and the Evolution of Politics." *Victorian Studies* 2:159–77.

Himmelfarb, Gertrude. 1968. *Victorian Minds.* New York: Knopf.

―――. 1986. *Marriage and Morals among the Victorians.* New York: Knopf.

Hobbes, Thomas. 1651. *Leviathan, or The Matter, Forme, & Power of a Common-wealth, Ecclesiasticall and Civill.* Ed. C. B. MacPherson. 1968. New York: Penguin.

Hofstadter, Richard. 1955. *Darwinism in American Thought.* Revised ed. Boston: Beacon Press.

Hull, David. 1973. *Darwin and His Critics.* Chicago: University of Chicago Press.

Hume, David. 1739-40. *A Treatise of Human Nature.* Ed. L. A. Selby-Bigge. 2d. ed. 1978. Oxford: Oxford University Press.

―――. 1777. *Enquiries Concerning Human Understanding and Concerning the Principles of Morals.* Ed. L. A. Selby-Bigge. 3d ed. Revised P. H. Nidditch. 1975. Oxford: Oxford University Press.

Huxley, Julian. 1948. "Eugenics and Society." *Man in the Modern World: Selected Essays.* New York: New American Library.

Huxley, Julian, and Thomas H. Huxley. 1947. *Touchstone for Ethics: 1893–1943.* New York: Harper and Brothers.

Huxley, Leonard. 1900. *Life and Letters of Thomas Henry Huxley.* 2 vols. New York: Appleton.

Huxley, Thomas H. 1866. *Lessons in Elementary Physiology.* New ed. 1881. London: Macmillan.

―――. 1879. *The Crayfish: An Introduction to the Study of Zoology.* Reprint. 1974. Cambridge: MIT Press.

Huxley, Thomas H. 1891. Introduction to *The Revolutionary Spirit*, by Théodore Rocquain, trans. J. D. Hunting. London: Swan, Sonnenschein.

―――. 1892. "An Apologetic Irenicon." *The Fortnightly Review* 52:557–71.

―――. 1893. "Evolution and Ethics." The Second Romanes Lecture. London: Macmillan.

―――. 1893–94. *Collected Essays*. 9 vols. London: Macmillan.

―――. 1936. *T. H. Huxley's Diary of the Voyage of HMS Rattlesnake*. Ed. Julian Huxley. Garden City, New York: Doubleday.

Huxley, Thomas H. et al. 1877. "A Modern Symposium: The Influence upon Morality of a Decline in Religious Belief." *Nineteenth Century* 1:531–46.

Irvine, William. 1955. *Apes, Angels, and Victorians: The Story of Darwin, Huxley, and Evolution*. New York: McGraw-Hill.

Jones, Gareth Steadman. 1971. *Outcast London: A Study in the Relationship between Classes in Victorian Society*. Reprint. 1984. New York: Penguin.

Kevles, Daniel J. 1985. *In the Name of Eugenics: Genetics and the Uses of Human Heredity*. Reprint. 1986. Berkeley: University of California Press.

Kohn, David. 1980. "Theories to Work By: Rejected Theories, Reproduction, and Darwin's Path to Natural Selection." *Studies in the History of Biology* 4:67–170.

Kropotkin, Petr. 1914. *Mutual Aid: A Factor of Evolution*. Reprint. 1955. Boston: Extending Horizons Books.

Landau, Misia. 1984. "Human Evolution as Narrative." *American Scientist* 72:262–68.

Lankester, E. Ray. 1880. *Degeneration: A Chapter in Darwinism*. London: Macmillan.

―――. 1907. *The Kingdom of Man*. London: Archibald Constable.

Leiss, William. 1972. *The Domination of Nature*. New York: Braziller.

Lightman, Bernard. 1987. *The Origins of Agnosticism: Victorian Unbelief and the Limits of Knowledge*. Baltimore: Johns Hopkins University Press.

Lovejoy, Arthur O. 1936. *The Great Chain of Being*. Cambridge: Harvard University Press.

―――. 1948. *Essays in the History of Ideas*. New York: G. P. Putnam's Sons.

Lubbock, John. 1870. *The Origins of Civilization and the Primitive Condition of Man*. London: Longmans, Green, and Company.

Malthus, Thomas R. 1798. *An Essay on the Principle of Population as it Affects the Future Improvement of Society and An Essay on the Principle of Population . . . and A summary View of the Principle of Population*. Ed. Anthony Flew. 1970. New York: Penguin.

Manier, Edward. 1978. *The Young Darwin and His Cultural Circle*. Dordrecht: Reidel.

Marsh, George Perkins. 1864. *Man and Nature; or, Physical Geography as Modified by Human Action*, ed. D. Lowenthal. Reprint. 1965. Cambridge: Harvard University Press.

Meyers, Greg. 1985. "Nineteenth Century Popularizations of Thermodynamics and the Rhetoric of Social Prophecy." *Victorian Studies* 29:35–66.

Mill, John Stuart. 1969. *Essays on Ethics, Religion, and Society.* Vol 10 of *Collected Works of John Stuart Mill.* Ed. J. M. Robson. Toronto: University of Toronto Press.

Mivart, St. George. 1871. "Darwin's *Descent of Man.*" In *Darwin and His Critics.* Ed. David Hull. 1973. Chicago: University of Chicago Press.

———. 1893. "Evolution in Professor Huxley." *Popular Science Monthly* 44:319–33.

Moore, George E. 1903. *Principia Ethica.* Reprint. 1984. Cambridge: Cambridge University Press.

Moore, James R. 1979. *The Post-Darwinian Controversies: A Study of the Protestant Struggle to Come to Terms with Darwin in Great Britain and America, 1870–1900.* Cambridge: Cambridge University Press.

Morgan, C. Lloyd. 1891. *Animal Life and Intelligence.* London: Edward Arnold.

———. 1896. *Habit and Instinct.* London: Edward Arnold.

———. 1923. *Emergent Evolution.* New York: Henry Holt.

———. 1933. *The Emergence of Novelty.* London: Williams and Norgate.

Noland, Richard. "T. H. Huxley on Culture." *The Personalist.* 45:94–111.

Ospovat, Dov. 1981. *The Development of Darwin's Theory: Natural History, Natural Theology, and Natural Selection, 1838–1859.* Cambridge; Cambridge University Press.

Paradis, James. 1978. *T. H. Huxley: Man's Place in Nature.* Lincoln: University of Nebraska Press.

Pearson, Karl. 1907. "On the Scope and Importance to the State of the Science of National Eugenics." 2d ed. 1909. London: Dulau and Company.

Quételet, Lambert Adolphe. 1835. *A Treatise on Man and the Development of His Faculties.* Trans. R. Knox. 1842. Reprint. 1968. New York: Burt Franklin.

Reid, Sir Wemyss, ed. 1899. *The Life of William Ewart Gladstone.* 2 vols. New York: G. P. Putnam's Sons.

Richards, Robert. 1981. "Instinct and Intelligence in British Natural Theology: Some Contributions to Darwin's Theory of the Evolution of Behavior." *Journal of the History of Biology* 14:193–230.

———. 1982. "Darwin and the Biologizing of Moral Behavior." In *The Prob-

lematic Science: Psychology in Nineteenth-Century Thought, ed. William R. Woodward and Mitchell G. Ash, 43–64. New York: Praeger.

———. 1987. *Darwin and the Emergence of Evolutionary Theories of Mind and Behavior* Chicago: University of Chicago Press.

Ritchie, D. G. 1891. *Darwinism and Politics*. 2nd ed. London: Swan Sonnenschein.

Ritvo, Harriet. 1987. *The Animal Estate: The English and Other Creatures*. Cambridge: Harvard University Press.

Ruse, Michael. 1979. *The Darwinian Revolution: Science Red in Tooth and Claw*. Chicago: University of Chicago Press.

———. 1986. *Taking Darwin Seriously: A Naturalistic Approach to Philosophy*. New York: Blackwell.

Sawyer, Paul. 1985. "Ruskin and Tyndall: The Poetry of Matter and the Poetry of Spirit." In *Victorian Science and Victorian Values: Literary Perspectives*, ed. James Paradis and Thomas Postlewait, 217–46. New Brunswick: Rutgers University Press.

Schwartz, Benjamin. 1964. *In Search of Wealth and Power: Yen Fu and the West*. Cambridge: Harvard University Press.

Schweber, S. S. 1977. "The Origin of the *Origin* Revisited." *Journal of the History of Biology* 10:229–316.

Smith, Adam. 1759. *The Theory of Moral Sentiments*. Reprint. 1976. Oxford: Oxford University Press.

———. 1776. *An Inquiry into the Nature and Causes of the Wealth of Nations*. Reprint. 1976. 2 vols. Oxford: Oxford University Press.

Spencer, Herbert. 1851. *Social Statics*. Revised ed. 1892. New York: Appleton.

———. 1893. "Evolutionary Ethics." *The Atheneum*, no. 3432:193–94.

———. 1897. *The Principles of Ethics*. American ed. 2 vols. Reprint. 1978. Indianapolis: Liberty Classics Press.

———. 1904. *Essays: Scientific, Political, and Speculative*. 3 vols. New York: Appleton.

Stanley, Oma. 1957. "T. H. Huxley's Treatment of Nature." *Journal of the History of Ideas* 18:120–27.

Stent, Gunther S., ed. 1980. "Introduction." In *Mortality as a Biological Phenomenon: The Presuppositions of Sociobiological Research*. Berkeley: University of California Press.

Stephen, Leslie. 1893. "Ethics and the Struggle for Existence." *The Contemporary Review* 64:157–70.

———. 1895. "The Huxley Memorial." *Nature Magazine* 53:183–86.

Stocking, George, W., Jr. 1971. "What's in a Name? The Origins of the Royal Anthropological Institute (1837–71)." *Man* 6:369–91.

————. 1987. *Victorian Anthropology*. New York: Free Press.

Tillyard, E.M.W. 1942. *The Elizabethan World Picture*. New York: Random House.

Thomas, Keith. 1983. *Man and the Natural World*. New York: Pantheon Books.

Toulmin, Stephen. 1982. *The Return to Cosmology: Postmodern Science and the Theology of Nature*. Berkeley: University of California Press.

Toynbee, Arnold. 1884. *The Industrial Revolution*. Reprint. 1956. Boston: Beacon Press.

Turner, Frank. 1974. *Between Science and Religion: The Reaction to Scientific Naturalism in Late Victorian England*. New Haven: Yale University Press.

Tylor, Edward. 1871. *Primitive Culture*. 2 vols. London: John Murray.

Wallace, Alfred. 1864. "The Origin of Human Races and the Antiquity of Man Deduced from the Theory of Natural Selection." *Journal of the Anthropological Society* 2:58–88.

————. 1900. "The Problem of Utility: Are Specific Characters Always or Generally Useful?" In *Studies, Scientific and Social*. 2 vols. London: Macmillan.

Whitehead, Alfred North. 1925. *Science and the Modern World*. New York: New American Library.

Willey, Basil. 1940. *The Eighteenth-Century Background: Studies on the Idea of Nature in the Thought of the Period*. Boston: Beacon Press.

Williams, Raymond. 1980. *Problems in Materialism and Culture: Selected Essays*. London: Verso.

Worster, Donald. 1977. *Nature's Economy: A History of Ecological Ideas*. Cambridge: Cambridge University Press.

Young, Robert. 1985. *Darwin's Metaphor: Nature's Place in Victorian Culture*. Cambridge: Cambridge University Press.

Bibliography II:
A Sociobiological Expansion

Abele, Lawrence G. and Sandra Gilchrist. 1977. "Homosexual Rape and Sexual Selection in Acanthocephalan Worms." *Science* 197:81–83.

Alexander, Richard D. 1979. *Darwinism and Human Affairs*. Seattle: University of Washington Press.

———. 1987. *The Biology of Moral Systems*. Chicago: Aldine de Gruyter.

Altman, Jeanne. 1980. *Baboon Mothers and Infants*. Cambridge: Harvard University Press.

Altman, Stuart A. 1962. "A Field Study of the Sociobiology of Rhesus Monkeys, *Macaca mullata*." *Annals of the New York Academy of Science* 102:338–435.

———. 1974. "Baboons, Space, Time, and Energy." *American Zoologist* 14:221–48.

Andersson, Malthe and M.O.G. Eriksson. 1982. "Nest Parasitism in Goldeneyes *Bucephala clangula*: Some Evolutionary Aspects." *American Naturalist* 120:1–16.

Andersson, Malthe, Christer G. Wiklund, and Helen Rundgren. 1980. "Parental Defense of Offspring: A Model and an Example." *Animal Behavior* 28:536–42.

Arak, Anthony. 1984. "Sneaky Breeders." In *Producers and Scroungers: Strategies of Exploitation and Parasitism*, ed. C. J. Barnard, 154–94. London: Croon Helm; New York: Chapman and Hall.

Baker, Myron Charles, and Stanley F. Fox. 1978. "Differential Survival in Common Grackles Sprayed with Turgitol." *American Naturalist* 112:675–82.

Barash, David P. 1976. "The Male Response to Apparent Female Adultery in the Mountain Bluebird *Sialia currucoides*: An Evolutionary Interpretation." *American Naturalist* 110:1097–1101.

———. 1977. "Sociobiology of Rape in Mallards (*Anas platyrhynchos*): Responses of the Mated Male." *Science* 197:788–89.

———. 1982. *Sociobiology and Behavior*. 2d ed. New York: Elsevier.

Beach, Frank A. 1964. "Biological Bases for Reproductive Behavior." In *Social Behavior and Organization among Vertebrates*, ed. W. Etkin 117–42. Chicago: University of Chicago Press.

———. 1978. "Sociobiology and Interspecific Comparisons of Behavior." In

Sociobiology and Human Nature, ed. M. S. Gregory, A. Silvers, and D. Sutch, 116–35. Chicago: University of Chicago Press.

Beecher, Michael D. and I. M. Beecher. 1979. "Sociobiology of Bank Swallows: Reproductive Strategy of the Male." *Science* 205:1282–85.

Belwood, Jacqueline J. and Glen K. Morris. 1987. "Bat Predation and its Influence on Calling Behavior in Neotropical Katydids." *Science* 238:64–67.

Berry, James F. and Richard Shine. 1980. "Sexual Size Dimorphism and Sexual Selection in Turtles (Order Testudines)." *Oecologia* 44:185–91.

Bierman, Gloria C. and Raleigh J. Robertson. 1981. "An Increase in Parental Investment during the Breeding Season." *Animal Behavior* 29:487–89.

Bigelow, Henry B. and William C. Schroeder. 1953. *Fishes of the Gulf of Maine. Fishery Bulletin*, U. S., 53.

Blaffer Hrdy, Sarah. 1977a. "Infanticide as a Primate Reproductive Strategy." *American Scientist* 65:40–55.

———. 1977b. *The Langurs of Abu*. Cambridge: Harvard University Press.

———. 1979. "Understanding Sociobiology." Interview by Georgia Litwack. *Boston Sunday Globe Magazine*, 8 April, 57.

Blaffer Hrdy, Sarah, and Glen Hausfater. 1984. "Comparative and Evolutionary Perspectives on Infanticide: Introduction and overview." In Hausfater and Blaffer Hrdy 1984, xiii–xxxv.

Blaustein, Andrew. 1983. "Kin Recognition Mechanisms: Phenotypic Matching or Recognition Alleles?" *American Naturalist* 121:749–54.

Bourke, Andrew F. G. 1988. "Worker Reproduction in the Higher Hymenoptera." *Quarterly Review of Biology* 64 (in press).

Bowden, Bradley S. 1969. "A New Method for Obtaining Precisely Timed Inseminations in Viviparous Fishes." *Progressive Fish Culturist* 31:229–30.

Bugos, Paul E. and Loraine M. McCarthy. 1984. "Ayoreo Infanticide: A Case Study." In Hausfater and Blaffer Hrdy 1984, 503–20.

Burger, Joanna. 1981. "On Becoming Independent in Herring Gulls: Parent-Young Conflict." *American Naturalist* 117:444–56.

Burger, Joanna, and C. G. Beer. 1975. "Territoriality in the Laughing Gull (*L. atricilla*)." *Behaviour* 55:301–20.

Burk, Theodore. 1982. "Evolutionary Significance of Predation on Sexually Signalling Males." *Florida Entomologist* 65:90–104.

Buskirk, Ruth, Cliff Frolich, and K. G. Ross. 1984. "The Natural Selection of Sexual Cannibalism." *American Naturalist* 123:612–25.

Campbell, Donald T. 1975. "On the Conflicts between Biological and Social

Evolution and Between Psychology and Moral Tradition." *American Psychologist* 117:1103–26.

———. 1978. "Social Morality Norms as Evidence of Conflict between Biological Human Nature and Social System Requirements." *Dahlem Konferenzen Life Sciences Research Report* 9:75–92.

Cavalli-Sforza, L. L. and Marcus W. Feldman. 1981. *Cultural Transmission and Evolution: A Quantitative Approach.* Princeton: Princeton University Press.

Chagnon, Napoleon. 1988. "Life Histories, Blood Revenge, and Warfare in a Tribal Population." *Science* 239:985–92.

Cheng, K. M., J. T. Burns, and F. McKinney. 1982. "Forced Copulation in Captive Mallards (*Anas platyrhynchos*), ii: Temporal Factors." *Animal Behavior* 30:695–99.

Clutton-Brock, T. H., S. D. Albon, R. M. Gibson, and F. E. Guinness. 1979. "The Logical Stag: Adaptive Aspects of Fighting in Red Deer (*Corvus elaphus* L.)." *Animal Behavior* 27:211–25.

Crump, Martha L. 1983. "Opportunistic Cannibalism by Amphibian Larvae in Temporary Aquatic Environments." *American Naturalist* 121:281–89.

Cuff, Wilfred R. 1980. "Behavioral Aspects of Cannibalism in Larval Walleye, *Stizostedion vitreum*." *Canadian Journal of Zoology* 58:1504–07.

Daly, Martin. 1978. "The Cost of Mating." *American Naturalist* 112:771–74.

Daly, Martin, and Margo Wilson. 1983. *Sex, Evolution, and Behavior.* Boston: Willard Grant Press.

———. 1988. *Homicide.* New York: Aldine de Gruyter.

Darwin, Charles. 1871. *The Descent of Man and Selection in Relation to Sex.* 2 vols. New York: Appleton.

Dawkins, Richard. 1976. *The Selfish Gene.* Oxford: Oxford University Press.

———. 1982. *The Extended Phenotype.* Oxford: W. H. Freeman and Company.

Day, Gordon M. 1953. "The Indian as an Ecological Factor in the Northeastern Forest." *Ecology* 34:329–46.

DeMartini, Edward F. 1987. "Paternal Defense, Cannibalism and Polygamy: Factors Influencing the Reproductive Success of Painted Greenling (Pisces, Hexagrammidae)." *Animal Behavior* 35:1145–58.

Dickeman, Mildred. 1979. "Female Infanticide, Reproductive Strategies and Social Stratification: A Preliminary Model." In *Evolutionary Biology and Human Social Behavior,* ed. N. Chagnon and W. Irons, 321–67. North, Scituate, Mass.: Duxbury.

Dillard, Annie. 1974. *Pilgrim at Tinker Creek.* New York: Harper's Magazine Press.

Dominey, Wallace J. 1980. "Female Mimicry in Male Bluegill Sunfish—A Genetic Polymorphism?" *Nature* 284:546–48.

Downhower, Jerry F. and Luther Brown. 1981. "The Timing of Reproduction and its Behavioral Consequences for Mottled Sculpins, *Cottus bairdi.*" In *Natural Selection and Social Behavior*, ed. R. D. Alexander and D. W. Tinkle, 78–95. New York and Concord: Chiron Press.

Dunford, Christopher. 1977. "Kin Selection for Ground Squirrel Alarm Calls." *American Naturalist* 111:782–85.

Ehrman, Lee. 1964. "Courtship and Mating Behavior as a Reproductive Isolating Mechanism in *Drosophila.*" *American Zoologist* 4:147–53.

Eichwort, Kathleen R. 1973. "Cannibalism and Kin Selection in *Labidomera clivicollis* (Coleoptera: Chrysomelidae)." *American Naturalist* 107:452–53.

Elwood, Robert W. and Malcolm C. Ostermeyer. 1984. "Does Copulation Inhibit Infanticide in Male Rodents?" *Animal Behavior* 32:293–94.

Endler, John A. 1978. "A Predator's View of Animal Color Patterns." *Evolutionary Biology* 11:319–64.

Estep, Daniel Q. and Katherine E. M. Bruce. 1981. "The Concept of Rape in Non-Humans: A Critique." *Animal Behavior* 29:1272–73.

Farr, James A. 1975. "The Role of Predation in the Evolution of Social Behavior of Natural Populations of the Guppy, *Poecilia reticulata* (Pisces: Poeciliidae)." *Evolution* 29:151–58.

———. 1980. "The Effects of Sexual Experience and Female Receptivity on Courtship-Rape Decisions in Male Guppies, *Poecilia reticulata* (Pisces: Poeciliidae)." *Animal Behavior* 28:1195–1201.

Forester, Don C. and David V. Lykens. 1986. "Significance of Satellite Males in a Population of Spring Peepers (*Hyla crucifer*)." *Copeia* 1986:719–24.

Fridgeirsson, Eyolfur. 1976. "Observations on Spawning Behaviour and Embryonic Development of the Icelandic Capelin." *Rit Fiskideildar* 5(no. 4):1–18.

Gargett, Valerie. 1978. "Sibling Aggression in the Black Eagle of the Matapos, Rhodesia." *Ostrich* 49:57–63.

Gould, Stephen Jay. 1980. "Sociobiology and the Theory of Natural Selection." *American Association for the Advancement of Science, Selected Symposium* 35:257–69.

———. 1982. "Darwinism and the Expansion of Evolutionary Theory." *Science* 216:380–87.

Gowaty, Patricia Adair. 1981. "Aggression of Breeding Eastern Bluebirds (*Sialia sialis*) toward their Mates and Models of Intra- and Interspecific Intruders." *Animal Behavior* 29:1013–27.

Gowaty, Patricia Adair. 1982. "Sexual Terms in Sociobiology: Emotionally Evocative and, Paradoxically, Jargon." *Animal Behavior* 30:630–31.

Gross, Mart R. 1979. "Cuckoldry in Sunfishes (*Lepomis*: Centrarchidae)." *Canadian Journal of Zoology* 57:1507–09.

Grunbaum, Adolf. 1952. "Causality and the Science of Human Behavior." *American Scientist* 40:665–76.

Guthrie, Daniel A. 1971. "Primitive Man's Relationship to Nature." *Bio-Science* 21:721–23.

Hamilton, William D. 1964. "The Genetical Evolution of Social Behavior." Parts 1, 2. *Journal of Theoretical Biology* 7:1–52.

———. 1967. "Extraordinary Sex Ratios." *Science* 156:477–88.

———. 1971. "Geometry for the Selfish Herd." *Journal of Theoretical Biology* 31:295–311.

Hamner, William M., P. P. Hamner, S. W. Strand, and R. W. Gilmer. 1983. "Behavior of Antarctic Krill, *Euphausia Superba*: Chemoreception, Feeding, Schooling, and Molting." *Science* 220:433–35.

Hausfater, Glen, and Sarah Blaffer Hrdy, eds. 1984. *Infanticide: Comparative and Evolutionary Perspectives*. New York: Aldine.

Hershkovitz, Philip. 1977. *Living New World Monkeys (Platyrhini)*, vol. 1. Chicago: University of Chicago Press.

Hoogland, John L. 1985. "Infanticide in Prairie Dogs: Lactating Females Kill Offspring of Close Kin." *Science* 230:1037–40.

Howard, Richard D. 1978. "The Evolution of Mating Strategies in Bullfrogs, *Rana catesbiana*." *Evolution* 32:850–71.

Huxley, Leonard. 1900. *Life and Letters of Thomas Henry Huxley*. 2 vols. New York: Appleton.

Huxley, Thomas H. 1893–94. *Collected Essays*. 9 vols. London: Macmillan.

Janzen, Daniel H. 1977a. "A Note on Optimal Mate Selection by Plants." *American Naturalist* 111:365–71.

———. 1977b. "Why Fruits Rot, Seeds Mold, and Meat Spoils." *American Naturalist* 111:691–713.

Johnson, Gary R., Susan H. Ratwik, and Timothy J. Sawyer. 1988. "The Adaptive Significance of Kin Terms in Patriotic Speech." In *The Sociobiology of Ethnocentrism*, ed. Vernon Reynolds, Vincent Falger, and Ian Vine, 157–74. London: Croon Helm.

Kaufmann, John H. 1974. "The Ecology and Evolution of Social Organization in the Kangaroo Family (Macropodidae)." *American Zoologist* 14:51–62.

Keenleyside, Miles H. A. 1972. "Intraspecific Intrusions Into Nests of Spawning Longear Sunfish (Pisces: Centrarchidae)." *Copeia* 1972:272–78.

Kellam, B. A. 1974. "Infanticide in England in the Later Middle Ages." *History of Childhood Quarterly* 1:368–88.

Kruuk, Hans. 1976. "The Biological Function of Gulls' Attraction Towards Predators." *Animal Behavior* 24:146–53.

Lacy, Robert C. and Paul W. Sherman. 1983. "Kin Recognition by Phenotype Matching." *American Naturalist* 121:489–512.

Lande, Russell. 1980. "Sexual Dimorphism, Sexual Selection, and Adaptation in Polygenic Characters." *Evolution* 34:292–305.

Langer, W. L. 1974. Infanticide: A Historical Survey." *History of Childhood Quarterly* 1:353–67.

Lank, David B., Pierre Mineau, Robert F. Rockwell, and Fred Cooke. 1988. "Intraspecific Nest Parasitism and Extra-Pair Copulation in Lesser Snow Geese." *Animal Behavior* (in press).

Leacock, Eleanor. 1980. "Social Behavior, Biology and the Double Standard." *American Association for the Advancement of Science, Selected Symposium* 35:465–88.

LeBoeuf, Burney J. 1974. "Male-Male Competition and Reproductive Success in Elephant Seals." *American Zoologist* 14:163–76.

Lopreato, Joseph. 1981. "Toward a Theory of Genuine Altruism in *Homo sapiens.*" *Ethology and Sociobiology* 2:113–26.

Lucas, Jeffrey R. 1985. "Partial Prey Consumption by Antlion Larvae." *Animal Behavior* 33:945–58.

Lumpkin, Susan, K. Kessel, P. G. Zenone, and C. J. Erickson. 1982. "Proximity between the Sexes in Ring Doves: Social Bonds or Surveillance?" *Animal Behavior* 30:506–13.

Lumsden, Charles J. and Edward O. Wilson. 1981. *Genes, Mind, and Culture.* Cambridge: Harvard University Press.

McPhail, J. D. 1969. "Predation and the Evolution of a Stickleback (Gasterosteidae)." *Journal of the Fisheries Research Board of Canada* 26:3183–208.

Massey, Adrianne. 1977. "Agonistic Aids and Kinship in a Group of Pigtail Macaques." *Behavioral Ecology and Sociobiology* 2:31–40.

May, Margaret Talmadge. 1968. *Galen on the Usefulness of the Parts of the Body.* Ithaca: Cornell University Press.

Maynard Smith, John. 1964. "Group Selection and Kin Selection." *Nature* 201:1145–47.

———. 1977. "Parental Investment: A Prospective Analysis." *Animal Behavior* 25:1–9.

Mayr, Ernst, and Will Provine. 1980. *The Evolutionary Synthesis.* Cambridge: Harvard University Press.

Mineau, Pierre, and Fred Cooke. 1979. "Rape in the Lesser Snow Goose." *Behaviour* 70:280–91.

Mundinger, Paul C. 1980. "Animal Cultures and a General Theory of Cultural Evolution." *Ethology and Sociobiology* 1:183–223.

O'Neill, Kevin M. and Howard E. Evans. 1981. "Predation on Conspecific Males by Females of the Beewolf *Philanthus basilaris* Cresson (Hymenoptera: Sphecidae)." *Journal of the Kansas Entomological Society* 54:553–56.

Organ, James A. and Della J. Organ. 1968. "Courtship Behavior of the Red Salamander, *Pseudotriton ruber*." *Copeia* 1968:217–23.

Parker, Geoffrey A. 1974. "Courtship Persistence and Female-Guarding as Male Time Investment Strategies." *Behaviour* 48:157–84.

———. 1979. "Sexual Selection and Sexual Conflict." In *Sexual Selection and Reproductive Competition in Insects*, ed. M. S. Blum and N. A. Blum, 123–26. New York: Academic Press.

Parsons, Jasper. 1971. "Cannibalism in Herring Gulls." *British Birds* 64:528–37.

Perrill, S. A., H. C. Gerhardt, and R. Daniel. 1978. "Sexual Parasitism in the Green Tree Frog (*Hyla cineria*)." *Science* 200:1179–80.

Picman, Jaroslav. 1977. "Intraspecific Nest Destruction in the Long-billed Marsh Wren, *Telmatodytes palustris palustris*." *Canadian Journal of Zoology* 55:1997–2003.

Polansky, Norman A., Carolyn Hally, and Nancy F. Polansky. 1975. *Profile of Neglect*. Community Services Administration Publication SRS 76–23037. Washington, D.C.: Government Printing Office.

Polis, Gary A. 1981. "The Evolution and Dynamics of Intraspecific Predation." *Annual Review of Ecology and Systematics* 12:225–51.

Power, Harry W. 1980. "On Bluebird Cuckoldry and Human Adultery." *American Naturalist* 116:705–09.

———. 1981. "The Question of Altruism." *Sociobiology* 6:7–21.

Power, Harry W., and C.G.P. Doner. 1980. "Experiments on Cuckoldry in the Mountain Bluebird." *American Naturalist* 116:689–704.

Pratt, Harold L., Jr. 1979. "Reproduction in the Blue Shark, *Prionace glauca*." *Fishery Bulletin*, U. S., 77:445–70.

Queller, David C. 1983. "Kin Selection and Conflict in Seed Maturation." *Journal of Theoretical Biology* 100:153–72.

Roberts, Thomas A. and James J. Kennelly. 1980. "Variation in Promiscuity among Red-winged Blackbirds." *Wilson Bulletin* 92:110–12.

Rohwer, Sievert. 1978. "Parent Cannibalism of Offspring and Egg Raiding as a Courtship Strategy." *American Naturalist* 112:429–40.

Rowland, William J. 1979. "Stealing Fertilizations in the Fourspine Stickleback, *Apeltes quadracus*." *American Naturalist* 114;602–04.

Ruse, Michael. 1986. *Taking Darwin Seriously: A Naturalistic Approach to Philosophy.* New York: Blackwell.

Schaffer, Wiliam M. 1968. "Intraspecific Combat and the Evolution of the Caprini." *Evolution* 22:817–25.

Schaller, George B. 1972. *The Serengeti Lion.* Chicago: University of Chicago Press.

Shaw, George Bernard. 1921. *Back to Methuselah.* Reprint 1965. Baltimore: Penguin Books.

Sherman, Paul W. 1980. "The Limits of Ground Squirrel Nepotism." *American Association for the Advancement of Science, Selected Symposium* 35:505–44.

———. 1981. "Reproductive Competition and Infanticide in Belding's Ground Squirrels and Other Mammals." In *Natural Selection and Social Behavior*, ed. R. D. Alexander and D. W. Tinkle, 311–31. New York and Concord, Chiron Press.

Sih, Andrew. 1980. "Optimal Foraging: Partial Consumption of Prey." *American Naturalist* 116:281–90.

Silk, Joan B., C. B. Clark-Wheatley, P. S. Rodman, and Amy Samuels. 1981. "Differential Reproductive Success and Facultative Adjustment of Sex Ratios among Captive Female Bonnet Macaques (*Macaca radiata*)." *Animal Behavior* 29:1106–20.

Singer, Peter. 1981. *The Expanding Circle.* New York: Farrar, Straus, and Giroux.

Speth, John D. 1983. *Bison Kills and Bone Counts.* Chicago: University of Chicago Press.

Stent, Gunther S. 1978. "Introduction: The Limits of the Naturalistic Approach to Morality." *Dahlem Konferenzen Life Sciences Research Report* 9:13–21.

Stinson, Christopher H. 1979. "On the Selective Advantage of Fratricide in Raptors." *Evolution* 33:1219–25.

Symons, Donald. 1979. *The Evolution of Human Sexuality.* Oxford: Oxford University Press.

Trail, Pepper W., Stuart D. Strahl, Jerram L. Brown. 1981. "Infanticide in Relation to Individual and Flock Histories in a Communally Breeding Bird, the Mexican Jay (*Aphelocoma ultramarina*)." *American Naturalist* 118:72–82.

Trivers, Robert L. 1971. "The Evolution of Reciprocal Altruism." *Quarterly Review of Biology* 46:35–57.

———. 1972. "Parental Investment and Sexual Selection." In *Sexual Selec-*

tion and the Descent of Man, 1871–1971, ed. B. Campbell, 136–79. Chicago: Aldine.

———. 1974. "Parent-Offspring Conflict." *American Zoologist* 14:249–64.

———. 1985. *Social Evolution*. Menlo Park, Calif.: Benjamin/Cummings.

Tuttle, Merlin D. and Michael J. Ryan. 1981. "Bat Predation and the Evolution of Frog Vocalizations in the Neotropics." *Science* 214:677–78.

Verner, Jared. 1964. "Evolution of Polygamy in the Long-billed Marsh Wren." *Evolution* 18:252–61.

Verrell, Paul. 1982. "The Sexual Behavior of the Red-spotted Newt, *Notophthalmus viridescens*." *Animal Behavior* 30:1224–36.

Villa, Paola. 1986. "Cannibalism in the Neolithic." *Science* 233:431–37.

Wasser, Samuel K., ed. 1983. *Social Behavior of Female Vertebrates*. New York: Academic Press.

Weinrich, James D. 1983. "Homosexual Behavior in Animals: A New Review of Observations from the Wild, and their Relationship to Human Sexuality." In *Medical Sexology: The Third International Congress*, ed. R. Forleo and W. Pasini, 288–95. Littleton, Mass.: PSG Publishing Company.

Wilkinson, Paul F. and Christopher C. Shank. 1976. "Rutting-Fight Mortality among Musk Oxen on Banks Island, Northwest Territories, Canada." *Animal Behavior* 24:756–58.

Willson, Mary F. and Nancy Burley 1983. *Mate Choice in Plants*, Princeton: Princeton University Press.

Wilson, David Sloan. 1980. *The Natural Selection of Populations and Communities*. Menlo Park, Calif.: Benjamin/Cummings.

Wilson, Edward O. 1975. *Sociobiology*. Cambridge: Harvard University Press.

———. 1978. *On Human Nature*. Cambridge: Harvard University Press.

Wimsatt, William C. 1980. "Reductionistic Research Strategies and their Biases in the Units of Selection Controversy." In *Scientific Discovery: Case Studies*, ed. T. Nickles, 213–59. Dordrecht: Reidl Publishing Company.

Wolf, Larry L. and F. Gary Stiles. 1970. "Evolution of Pair Cooperation in a Tropical Hummingbird." *Evolution* 24:759–73.

Wolf, Nancy G. 1983. "Behavioral Ecology of Herbivorous Reef Fishes in Mixed-Species Foraging Groups." Ph.D. diss., Cornell University.

Index

237